Schriftenreihe der Institute für Systemdynamik (ISD) und optische Systeme (IOS)

Editors-in-Chief

Jürgen Freudenberger, Konstanz, Germany

Johannes Reuter, Konstanz, Germany

Matthias Franz, Konstanz, Germany

Georg Umlauf, Konstanz, Germany

Die „Schriftenreihe der Institute für Systemdynamik (ISD) und optische Systeme (IOS)" deckt ein breites Themenspektrum ab: von angewandter Informatik bis zu Ingenieurswissenschaften. Die Institute für Systemdynamik und optische Systeme bilden gemeinsam einen Forschungsschwerpunkt der Hochschule Konstanz. Die Forschungsprogramme der beiden Institute umfassen informations- und regelungstechnische Fragestellungen sowie kognitive und bildgebende Systeme. Das Bindeglied ist dabei der Systemgedanke mit systemtechnischer Herangehensweise und damit verbunden die Suche nach Methoden zur Lösung interdisziplinärer, komplexer Probleme. In der Schriftenreihe werden Forschungsergebnisse in Form von Dissertationen veröffentlicht.

The "Series of the institutes of System Dynamics (ISD) and Optical Systems (IOS)" covers a broad range of topics: from applied computer science to engineering. The institutes of System Dynamics and Optical Systems form a research focus of the HTWG Konstanz. The research programs of both institutes cover problems in information technology and control engineering as well as cognitive and imaging systems. The connective link is the system concept and the systems engineering approach, i.e. the search for methods and solutions of interdisciplinary, complex problems. The series publishes research results in the form of dissertations.

More information about this series at http://www.springer.com/series/16265

Pascal Laube

Machine Learning Methods for Reverse Engineering of Defective Structured Surfaces

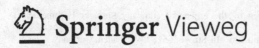
Springer Vieweg

Pascal Laube
Konstanz, Germany

Dissertation Universität Konstanz, 2019, u.d.T.: Pascal Laube "Machine Learning Methods for Reverse Engineering of Defective Structured Surfaces" This research has been partly funded by the Federal Ministry of Education and Research (BMBF) of Germany (pn 02P14A035).

ISSN 2661-8087 ISSN 2661-8095 (electronic)
Schriftenreihe der Institute für Systemdynamik (ISD) und optische Systeme (IOS)
ISBN 978-3-658-29016-0 ISBN 978-3-658-29017-7 (eBook)
https://doi.org/10.1007/978-3-658-29017-7

This Springer Vieweg imprint is published by the registered company Springer Fachmedien Wiesbaden GmbH part of Springer Nature.
The registered company address is: Abraham-Lincoln-Str. 46, 65189 Wiesbaden, Germany

To Jennifer and Levin

Danksagung

Als ich im August 2013 am Institut für Optische Systeme die Stelle des Institutsreferenten übernahm wusste ich wenig über geometrisches Modellieren, maschinelles Lernen oder darüber, wie spannend, interessant und lehrreich die folgenden Jahre werden würden. In dieser Zeit begleiteten mich Mentoren, Kollegen, Studenten sowie geliebte Menschen aus meinem persönlichen Umfeld denen ich zu großem Dank verpflichtet bin. Zuerst möchte ich meinem Hauptbetreuer und Mentor Prof. Dr. Georg Umlauf danken. Ohne mich näher zu kennen, gab er mir 2013 die Chance mich auf einem Feld zu beweisen, welches mir weitgehend unbekannt war. In der Zeit der Promotion war er immer mein zentraler Orientierungspunkt. Eine solch enge Zusammenarbeit verstehe ich nicht als selbstverständlich und ich bin ihm für seine Führung in dieser Zeit sehr dankbar. Seine Expertise, seine Klarheit und seine Fähigkeit komplexe Themen greifbar zu machen, zeichnen ihn als Professor und Mentor aus. Für seine Rücksicht und seine bestärkenden Worte in schwierigen Phasen, professionell wie privat, bin ich ihm sehr dankbar. Ohne ihn wäre diese Dissertation nicht möglich gewesen. Weiter möchte ich Prof. Dr. Matthias Franz für seine richtungsweisende Betreuung danken. Auf sein Gespür für mögliche neue Ansätze konnte ich mich immer verlassen. Sein Enthusiasmus hat mich häufig angesteckt. An der Universität Konstanz möchte ich mich bei Prof. Dr. Oliver Deussen bedanken. Die Gespräche mit ihm und seine Hilfestellungen führten mich maßgeblich durch die Prozesse der Universität. Ein besonderer Dank gilt meinen Kollegen am IOS Konstanz. Ich danke Michael Grunwald für die Zusammenarbeit bei zahlreichen Forschungsprojekten. Die gemeinsamen Gespräche eröffneten mir stets neue Perspektiven. Für neue Impulse und den regen Austausch möchte ich weiter Martin Schall, Matthias Herrmann und Klaus Denker danken. Bei Michael, Martin und Matthias möchte ich mich auch für die zahlreichen wichtigen Korrekturen und Anmerkungen zu dieser Dissertation bedanken. Auch bei Fabian Freiberg und Tobias Birkle möchte ich mich für die Korrekturen zu dieser Dissertation bedanken.

Ohne meine Frau Jennifer hätte ich den Schritt in die Promotion nie gewagt. Die Quelle aller positiven Energie, Bestärkung und Liebe. Ich entschuldige mich für die späten Stunden und bin ihr unendlich dankbar für ihr unerschöpfliches Verständnis. Ich bewundere Sie für ihren Realismus und ihre Gelassenheit. Diese Dissertation ist zu großen Teilen dein Verdienst – Danke. Für ihren unbeugsamen Rückhalt möchte ich meinen Eltern Kerstin und Stephan danken. Beharrlich standen sie, auch bei starkem Gegenwind, hinter mir. Vielen Dank für eure Unterstützung, eure Konsequenz und euren Zuspruch. Meinen Schwiegereltern Regina und Erhard möchte ich für die bedingungslose Unterstützung in allen Lebenssituationen danken. Vielen Dank, dass ihr mir unter die Arme greift, wenn es nötig ist. Eine wichtige Stütze über den gesamten Schaffensprozess waren Cinzia und Björn. Danke, dass ihr ein Teil meines Lebens seid.

Abstract

Reverse engineering of 3d-scanned functional surfaces is an essential task of modern manufacturing processes. Since production is highly automated, reverse engineering needs to become independent from user interaction, while maintaining high-quality results. Many functional surfaces like, e.g. injection moulds for the production of parts, consist of a base geometry and detailed surface structure which further complicates the reconstruction of a CAD model. Due to wear and tear of such surfaces the digital repair of defects is an essential part of the reconstruction pipeline. Recent achievements of machine learning methods in the field of computer vision are remarkable, but successful applications to CAD problems are still rare.

This doctoral thesis presents machine learning approaches for three key problems of reverse engineering of defective structured surfaces: First, the approximation of parametric curves and surfaces, which is essential to the reconstruction of CAD models, requires the computation of parametrizations. The parametrization methods presented in this thesis demonstrate that the prediction of parametrizations using machine learning can produce tight approximations. Second, parts to be produced often consist of patches of geometric primitives. The proposed machine learning approaches for the classification of geometric primitives in point clouds simplify the step of fitting primitives to patches. Third, the surface structure of functional surfaces is usually represented as high-resolution grayscale texture. Thus, the digital repair of the surface structure becomes an image inpainting problem. The application of neural texture synthesis for the inpainting of high-resolution textures are investigated in this thesis.

The proposed methods aim to improve the reconstruction quality while further automating the process. The contributions demonstrate that machine learning can be a viable part of the reverse engineering pipeline.

Contents

List of Abbreviations

CAD	Computer Aided Design
CNN	Convolutional Neural Network
CSG	Constructive Solid Geometry
DPKP	Dominant-Point-based Knot Placement method by Park et al. [71]
EI	Exemplar-based Inpainting method by Criminisi et al. [15]
HD	Hausdorff Distance
KSN	Knot Selection Network
LCM	Local Curvature Maximum points
MLP	Multilayer Perceptron
MSE	Mean Squared Error
NKTP	Knot placement method by Piegl and Tiller [73]
NN	Neural Network
NURBS	Non-Uniform Rational B-Spline
PARNET	Parametrization using Neural Networks (Section 3.3)
PITS	Patch-based Inpainting by Texture Synthesis (Chapter 5)
PPN	Point Parametrization Network
PSCC17	Content-aware fill method of Photoshop CC17
RANSAC	Random Sample Consensus
RBF	Radial Basis Function
RGB	Red, Green, and Blue color model
SKP	Score-based Knot Placement (Section 3.2)
SVM	Support Vector Machine
TPR	True Positive Rate

1 Introduction

Modern functional plastic surfaces are usually produced using injection moulds. In plastic injection moulding, the respective material is plasticized and injected under pressure into an injection mould made of hardened steel. After hardening to a solid state, the moulded part can be removed from the injection mould. The *basic geometry* of the inside of the injection mould together with the *surface structure* is known as the cavity. The cavity is transferred to the finished component. Well-known examples which are produced this way are, e.g. various parts of the interior of modern vehicles such as the dashboard, but also toys such as Lego bricks or Playmobil figures. Figure 1.1 shows an example of an injection mould for the production of airbag cases.

Since the visual quality of parts produced by injection moulding depends to a large extent on the quality of the injection mould, it has to be in perfect condition. Local defects of the injection mould may occur due to wear and tear as well as operating errors. These defects will be transferred to the moulded part. Due to the high purchase value of the moulds, repairing them is preferred over replacement. The situation is aggravated by the fact that the original CAD is often incomplete or not available. The steps of the repair process are mainly manual and the repair quality result is judged by eye. A systematic, partially automated solution for this process does not yet exist.

A prototype of a system for the laser-based repair of structured mould inserts, which is currently developed in the project *ToolRep*, is planned to carry out this repair semi-automatically. The prototype will be equipped with two surface scanners for coarse and detail capture of

© Springer Fachmedien Wiesbaden GmbH, part of Springer Nature 2020
P. Laube, *Machine Learning Methods for Reverse Engineering of Defective Structured Surfaces*, Schriftenreihe der Institute für Systemdynamik (ISD) und optische Systeme (IOS), https://doi.org/10.1007/978-3-658-29017-7_1

Figure 1.1: Airbag case injection mould with leather grain structure.
Source: Fraunhofer IPT.

the cavity. In this work we assume that the coarse scan is performed using some form of laser-line scanner and details are captured using coherence scanning interferometry. Both scanning methods result in three-dimensional point clouds of the cavity. After *reverse engineering* of the cavitiy, which includes digitally repairing the defective region, high-precision laser structuring restores the surface. Figure 1.2 shows the individual steps of the physical repair chain.

To represent functional surfaces with detailed surface structure two different techniques are combined: The base geometry, in the form of a computer-aided design (CAD) part, and an accompanying two-dimensional heightmap of the surface structure.

Most functional surfaces are designed by one of two approaches:

Constructive Solid Geometry (CSG) Objects are modelled by applying boolean operations such as union or intersection to solid geometric primitives like planes, spheres or tori.

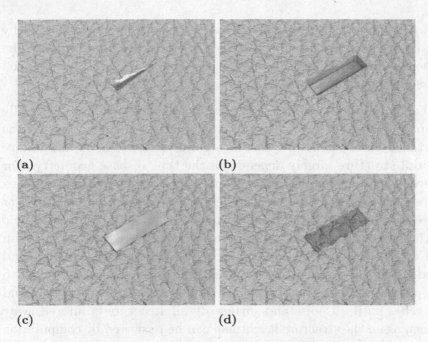

(a) (b)

(c) (d)

Figure 1.2: Consecutive steps (a)-(d) of the automated physical repair of
a structured surface. (a) Closeup on a structured injection
mould cavity with a defect. (b) Excavation of the defective
region. (c) After the application of a filler metal.
(d) Final repair by laser induced surface structuring. Source:
Fraunhofer IPT.

Freeform Surfaces Objects are modelled by definition of so-called
control-points to define the shape of a set of freeform surfaces. Most
operations like fairing, or smoothing are based on the manipulation
of single control points. A well-known freeform surface representation
are the so-called *non-uniform rational B-splines* (NURBS).

While being used separately in the past, modern CAD systems mix
both approaches.

The detailed surface structure is represented by, often tileable, grayscale
heightmaps. In contrast, to bump mapping, these heightmaps are

used as an actual displacement on the base geometry. This is why in computer graphics they are also called displacement maps. The specific displacement values, as well as the correspondence between base geometry and surface structure, is up to the designer.

The goal of the reverse engineering process in ToolRep is a digitally repaired representations of the cavity's base geometry as well as detailed surface structure. How to separate base geometry and surface detail structure largely depends on the type of base geometry. For reverse engineering, the point cloud is first segmented into patches. These patches are then approximated by either geometric primitives or freeform surfaces. If the base geometry is representable by geometric primitives, the detailed structure may be recovered by computation of distances between the geometric primitive and the point cloud patch. For the approximation of freeform surfaces, we propose first to apply surface relaxation methods [40, 83] to triangulated point cloud patches until all noise and surface detail structure is filtered away. Then again the structure heightmap can be recovered by computation of the distance between the approximation of the relaxed and the original point cloud.

Assuming, points that correspond to the surface defect have been removed from the point cloud, the base geometry can be considered repaired after approximation. Repair of the detailed surface structure on the other hand needs to be handled in grayscale image space.

1.1 Machine Learning for Reverse Engineering

In order to guarantee a mostly autonomous repair process, reverse engineering of the scanned surface needs to be independent of user interaction while maintaining high-quality results. Recent success using machine learning methods to solve computer vision related problems are remarkable but successful applications to CAD problems, particularly in reverse engineering, are still rare. The proposed thesis aims to develop machine learning methods for reverse engineering of defective structured surfaces that improve reconstruction quality while further automating the process.

Three key problems have been identified. The solution of these three key problems will further contribute to automating reverse engineering of structured surfaces:

- Approximation of point clouds by NURBS or T-Spline surfaces requires the preceding computation of a parametrization. This parametrization has a large impact on the resulting approximation quality. We propose the application of support vector machines and neural networks to compute parametrizations for B-spline curve and T-spline surface approximation.

- The classification of point clouds with the correct geometric primitive is usually implicit. Different primitive types are fitted, and based on error thresholds the best matching primitive is selected. This approach results in frequent misclassification and long computation times. We propose to apply machine learning, specifically support vector machines, to the problem of geometric primitive classification.

- Since the detailed surface structure is represented as grayscale texture, filling in the defective region becomes an image inpainting problem. To prevent the loss of structural information, the resulting heightmaps have to be of high-resolution. Recently, convolutional neural networks (CNN) have been applied successfully to the problems of texture synthesis and image inpainting. Due to limited computational resources, processing high-resolution images with neural networks is still an open problem. We propose to inpaint missing regions by two-scale resolution CNN texture synthesis.

In Figure 1.3 a schematic overview of reverse engineering of structured surfaces, and the integration of this thesis in the context of the above mentioned three key problems is given.

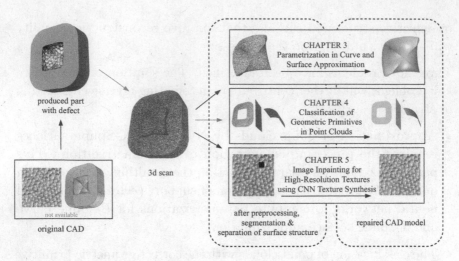

Figure 1.3: Schematic overview of the reverse engineering process and the integration of the key contributions of this thesis.

1.2 Organisation of this Thesis

Beginning with relevant fundamentals in Chapter 2, an introduction to machine learning (Section 2.1), as well as parametric curves and surfaces (Section 2.2) is given.

After this introductory chapter the organisation of this thesis is based on the previously described key problems.

In Chapter 3, approaches for the computation of parametrizations for curve and surface approximation are presented. Section 3.2 covers the computation of knot vectors using support vector machines. In Section 3.3 the prediction of parametric values as well as knots is approached by applying interdependent neural networks. The different methods are then evaluated in the context of T-spline surface approximation in Section 3.4.

Chapter 4 evaluates the performance of different geometric features and their combinations when used for the classification of geometric primitives by support vector machines. Results are compared to neural networks using a volumetric representation.

Chapter 5 covers the application of neural texture synthesis for the inpainting of high-resolution color images as well as heightmaps.

Each of the chapters 3 to 5 includes a section listing related works. The decision for separate related works sections instead of a single, global one is based on their need for an introduction, which is given before each section. While each of the chapters contains an independent conclusion, an overall conclusion is given in Chapter 6. With the exception of sections 1.3, 6 and this section, the first-person plural personal pronoun is applied since the presented contributions were made in collaboration. The singular form is only applied when comparing the contributions of different authors or if the context reflects my own opinion.

1.3 Publications

During my time working on the topic of reverse engineering I published various papers that reflect the results obtained throughout this period. This section lists all publications, this thesis is based upon. I will outline the contributions made by each author. The order of the publication list is based upon the order of the chapters of this thesis.

- [54] P. Laube, M. O. Franz, & G. Umlauf. **Learnt Knot Placement in B-spline Curve Approximation using Support Vector Machines**. Computer Aided Geometric Design, 62 (2018), pp. 104-116. - *(Chapter 3 of this thesis)*
  In this work, we present a method to compute knot vectors for B-Spline curve approximation using Support Vector Machines. The inspiration for this approach came from conversations with G. Umlauf, and our previous work using Support Vector Machines for classification of geometric primitives. I implemented and evaluated all methods that led to this publication. I wrote the paper in close

cooperation with G. Umlauf. M. O. Franz supervised the work regarding the application of machine learning and the corresponding sections of the paper.

- [53] P. Laube, M. O. Franz, & G. Umlauf. **Deep Learning Parametrization for B-Spline Curve Approximation**. International Conference on 3D Vision 3DV (2018), pp. 691-699. IEEE. - *(Chapter 3 of this thesis)*
  So far most approaches for parametrization in curve approximation regarded knot computation and the computation of parametric values as separate problems. In the listed paper we present interdependent neural networks to predict both the knot vector, and parametric values by including B-spline curve approximation into the neural network architecture. After discussions with G. Umlauf on how to become more independent of synthetic data sets, I decided to work on the integration of B-Spline curve approximation into neural networks. I did all implementations and evaluations. The publication was written in close cooperation with M. O. Franz and G. Umlauf.

- [9] M. Caputo, K. Denker, M. O. Franz, P. Laube, & G. Umlauf. **Support Vector Machines for Classification of Geometric Primitives in Point Clouds**. International Conference on Curves and Surfaces (2014), pp. 80-95. Springer. - *(Chapter 4 of this thesis)*
  In previous works by K. Denker geometric features together with hard thresholds had been used to classify geometries. Using machine learning for this classification task was the goal of the master's thesis by M. Caputo. Most of the implementations came from M. Caputo. I supported the work by implementing the methods for curvature estimation. Initial evaluations came from M. Caputo. I refined the initial evaluations and findings and wrote this publication in close cooperation with G. Umlauf. M. O. Franz supervised the application of machine learning methods and the writing of the corresponding sections of the paper.

- [8] M. Caputo, K. Denker, M. Franz, P. Laube, & G. Umlauf.

Learning Geometric Primitives in Point Clouds. Eurographics Symposium on Geometry Processing SGP (2014). Poster paper. Eurographics Association. - *(Chapter 4 of this thesis)*
To present the idea to use machine learning for primitive classification and to get feedback from the community, we published this poster paper. The ideas presented were then refined, and led to the above-described publication "Support Vector Machines for Classification of Geometric Primitives in Point Clouds". The poster and the paper were created by me and in cooperation with G. Umlauf. M. O. Franz reviewed the sections dealing with machine learning.

- [57] P. Laube, M. O. Franz, & G. Umlauf. **Evaluation of Features for SVM-based Classification of Geometric Primitives in Point Clouds**. Machine Vision Applications MVA (2017), pp. 59-62. IEEE. - *(Chapter 4 of this thesis)*
Our published works so far had been focused on the aspects of data generation and machine learning. In this publication, we present a missing in-depth analysis of known and novel geometric features and their combinations for geometric primitive classification. We also evaluated the impact of point cloud density of real and synthetic data for classification. On the basis of code by M. Caputo, I implemented new functionalities and did the evaluations. I wrote the paper in close cooperation with G. Umlauf. M. O. Franz reviewed the sections dealing with machine learning.

- [17] M. Danhof, T. Schneider, P. Laube, & G. Umlauf. **A Virtual-Reality 3d-Laser-Scan Simulation**. Symposium on Information and Communication Systems BW-CAR— SINCOM (2015), pp. 68-73. - *(Chapter 4 of this thesis)*
It became increasingly clear that the performance of machine learning methods for primitive classification were strongly influenced by the structure of the point clouds in the training data set. We decided to develop a simulation environment that imitated the functionality of a hand-held laser scanner in virtual reality. M. Danhof and T. Schneider developed this environment in the context of their bachelor's theses, and under my supervision. Based on

these theses I wrote the publication in close cooperation with G. Umlauf.

- [55] P. Laube, M. Grunwald, M. O. Franz, & G. Umlauf. **Image Inpainting for High-Resolution Textures using CNN Texture Synthesis**. EG UK Computer Graphics & Visual Computing (2018). EGUK. Best Short Paper Award. - *(Chapter 5 of this thesis)* In this publication, we present a method for inpainting high-resolution images using neural network texture synthesis. The idea for this patch-based two-scale resolution approach was inspired by previous work of M. Grunwald on texture synthesis and novelty detection. I implemented this inpainting approach on the basis of a texture synthesis prototype by M. Grunwald and T. Birkle. I wrote the paper in cooperation with M. O. Franz and under the supervision of G. Umlauf.

Furthermore, I worked on publications which are not part of this thesis but influenced the direction of my research:

- [36] M. Grunwald, M. Hermann, F. Freiberg, P. Laube, & M. O. Franz. **Optical surface inspection: A novelty detection approach based on CNN-encoded texture features**. Applications of Digital Image Processing, 10752 (2018). International Society for Optics and Photonics.

- [37] M. Grunwald, P. Laube, M. Schall, G. Umlauf, & M. O. Franz. **Radiometric calibration of digital cameras using neural networks**. Optics and Photonics for Information Processing XI, 10395 (2017). International Society for Optics and Photonics.

- [56] P. Laube, & G. Umlauf. **A short survey on recent methods for cage computation**. Symposium on Information and Communication Systems BW-CAR— SINCOM (2016), pp. 37-42.

- [38] M. Grunwald, J. Müller, M. Schall, P. Laube, G. Umlauf, & M. O. Franz. **Pixel-wise Hybrid Image Registration on Wood Decors**. Symposium on Information and Communication Systems BW-CAR— SINCOM (2015), pp. 24-29.

- [97] L. Thießen, P. Laube, M. O. Franz, & G. Umlauf. **Merging Multiple 3D Face Reconstructions**. Symposium on Information and Communication Systems BW-CAR— SINCOM (2014), pp. 68-73.

2 Fundamentals

2.1 Machine Learning Methods

In this thesis we will consider two main areas of application for machine learning namely *classification* and *regression*. In the so-called *training* phase the machine learning algorithms parameters need to be adapted with respect to the problem. Let us define a set of n training samples $\mathbf{x}_i$ with class labels y_i together called the training set. The goal of classification now is for the machine learning algorithm to decide on a class label for some previously unseen sample $\mathbf{x}$. Since the class labels of the training set are known beforehand, the training is called supervised. The performance of the trained machine learner is evaluated using some form of a performance measure. A simple performance measure for classification would be the number of correctly classified samples. Besides evaluation on the training set, performance is also evaluated for a set of samples that has not been used in the training phase called the validation set. While the performance on the training set indicates if the machine learner can adapt to the problem, its performance on the validation set hints at how well it generalizes to unseen data. If the corresponding labels y_i do not consist of class labels but continuous values this is regarded as a regression problem. While classification and regression are the two main fields of application of machine learning it is also applied for e.g. representation learning [3], data generation like the well-known generative adversarial nets [35] and to a multitude of other problems. Besides the actual adaption of the parameters of the machine learning algorithm, pre-processing is a vital part of many machine learning methods. While almost always including some form

© Springer Fachmedien Wiesbaden GmbH, part of Springer Nature 2020
P. Laube, *Machine Learning Methods for Reverse Engineering of Defective Structured Surfaces*, Schriftenreihe der Institute für Systemdynamik (ISD) und optische Systeme (IOS), https://doi.org/10.1007/978-3-658-29017-7_2

of input normalization, pre-processing often includes *feature extraction*. The goal of feature extraction is to generate features of, preferably lower dimensionality while preserving most of the task-specific relevant information.

The following sections cover two machine learning techniques, namely Support Vector Machines (SVM) and Neural Networks (NN) which will be used in this thesis.

2.1.1 Support Vector Machines

The support vector machine (SVM) is a maximum *margin* classifier. This name stems from its purpose to maximize the minimum distance of samples $\mathbf{x}_i$, with class labels $y_i \in \{-1, 1\}$, to a hyperplane that separates the two classes. SVMs are derived from the concept of separating a set of sample vectors $\mathbf{x}_i$ by a discriminant function. In the case of a linear model this discriminant has the form

$$f(\mathbf{x}) = \omega^T \phi(\mathbf{x}) + b, \tag{2.1}$$

where $\phi(\mathbf{x})$ is some feature-space transformation, a normal vector ω and a bias b. The prediction of class membership can be decided according to the sign of $f(\mathbf{x})$. For a separable problem there exist infinitely many hyperplanes that separate the two classes.

The SVM margin is defined as the smallest distance of an arbitrary sample from the separating hyperplane. See Figure 2.1 for a two dimensional linear SVM example. The goal is to find a hyperplane that separates the two distinct classes so that $y_i f(\mathbf{x}_i) > 0$ for all samples of the training set. Since the distance of an arbitrary sample $\mathbf{x}$ to the decision hyperplane, is given by $|f(\mathbf{x})|/||\omega||$ we can find a maximum margin solution to the classification problem by maximizing

$$\min_i \left\{ \frac{y_i(\omega^T \phi(\mathbf{x}_i) + b)}{||\omega||} \right\}$$

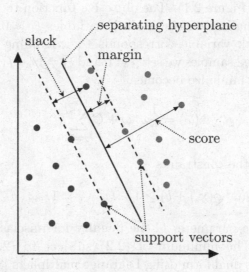

Figure 2.1: Separating hyperplane, margin, score, support vectors, and slack for a two dimensional feature space and two classes (red, blue).

with respect to ω and b. Since rescaling ω and b by a constant factor does not change the distance of $\mathbf{x}_i$ to the decision hyperplane we can set

$$y_i(\omega^T \phi(\mathbf{x}_i) + b) = 1$$

for the sample with minimal distance. This can be transformed to the problem of minimizing the quadratic function

$$\Phi(\omega) = \frac{1}{2}\omega^T \omega$$

with respect to the constraints

$$y_i(\omega^T \phi(\mathbf{x}_i) + b) \geq 1, \quad i = 1, \ldots, n.$$

For noisy data requiring linear separability might be too restrictive. Thus, *slack variables* ξ_i are used to allow for some data inside or beyond the margin (see Figure 2.1). The objective function to minimize now additionally penalizes excessive slack variables $\xi_i \geq 0$, $i = 1, \ldots, n$, where each slack variable corresponds to a training sample. For correctly classified samples we set $\xi_i = 0$ and $\xi_i = |y_i - f(\mathbf{x}_i)|$ otherwise. The function to minimize becomes

$$\Phi(\omega) = \frac{1}{2}\omega^T\omega + C\sum_{i=1}^{n}\xi_i \tag{2.2}$$

with respect to the constraints

$$y_i(\omega^T\phi(\mathbf{x}_i) + b) \geq 1 - \xi_i, \quad i = 1, \ldots, n. \tag{2.3}$$

By adapting the parameter C the penalty for misclassification can be adjusted. The minimum of (2.2) subject to (2.3) is usually computed via its dual form using Lagrange multipliers [39]. Then, the Karush-Kuhn-Tucker-conditions imply that the weight vector ω can be represented in terms of the Lagrange multipliers $\alpha = (\alpha_1, \ldots, \alpha_n)$ and the training data as $\omega = \sum_{i=1}^{n} \alpha_i\, y_i\, \phi(\mathbf{x}_i)$. So, instead of minimizing (2.2) the resulting Lagrangian

$$\Phi^*(\alpha) = \sum_{i=1}^{n}\alpha_i - \frac{1}{2}\sum_{i=1}^{n}\sum_{j=1}^{n}\alpha_i\alpha_j\, y_iy_j\, \phi(\mathbf{x}_i)^T\phi(\mathbf{x}_j). \tag{2.4}$$

is maximized with respect to the constraints

$$\sum_{i=1}^{n}\alpha_iy_i = 0, \qquad 0 \leq \alpha_i \leq C, \qquad i = 1, \ldots, n. \tag{2.5}$$

If $\alpha_i \neq 0$, the corresponding $\mathbf{x}_i$ is called *support vector*.

Substituting $\phi(\mathbf{x}_i)^T\phi(\mathbf{x}_j) = K(\mathbf{x}_i, \mathbf{x}_j)$ in (2.4) we are able to introduce a *kernel* function. By introducing kernel functions, we can project samples to a feature space of possibly infinite dimensions implicitly. Well known kernel functions include e.g. the Gaussian RBF kernel

$$K(\mathbf{x}_i, \mathbf{x}_j) = \exp(-\gamma\|\mathbf{x}_i - \mathbf{x}_j\|^2), \tag{2.6}$$

with the free kernel parameter γ. The classifier for new data $\mathbf{x}$ is given by $\text{sign}(m_{\alpha,b}(\mathbf{x}))$ with the *data margin*

$$m_{\alpha,b}(\mathbf{x}) = \sum_{i=1}^{n} \alpha_i y_i K(\mathbf{x}_i, \mathbf{x}) + b. \tag{2.7}$$

The *score* for predicting $\mathbf{x}$ in the positive class is $m_{\alpha,b}(\mathbf{x})$, while the score for prediction in the negative class is $-m_{\alpha,b}(\mathbf{x})$.

While SVMs can also be applied to regression problems, we solely use them for classification. For more on support vector regression the interested reader is referred to [16]. Since solving for α and b means optimizing a quadratic function with respect to the so called box constraints (2.5) training the SVM means solving a quadratic programming problem.

Model Selection

Finding parameters C and γ is referred to as *model selection*. This optimization is usually based on a non-uniform *grid search* in the two-dimensional (C, γ)-space. Thus, a trained SVM is completely determined by the SVM parameters (ω, b, C, γ).

To ensure independence of training and test data sets *k-fold cross-validation* is used. The original data set is partitioned into k distinct subsets of equal size, the so-called folds. Each fold is used as a test set once while the SVM is trained on all other folds. The accuracy of the model is then estimated by averaging over the k resulting accuracies. This way the impact of data set distribution on the computed model accuracy can be reduced which leads to improved performance predictions.

Multi-Class Classification

The SVM is a binary classifier. There exist different methods to enable multi-class classification of which the following two are the most popular:

One-Versus-All For One-Versus-All classification k different support vector machines are trained where k is the number of classes [101]. Each support vector machine is trained to distinguish one class from all others. The ith classifier is trained to classify the ith class as positive and the remaining $k - 1$ classes as negative. After training, the class membership is decided upon the support vector machine yielding the highest output value. This output value is proportional to the signed distance of the data point to the hyperplane.

One-Versus-One For One-Versus-One classification $k(k-1)/2$ SVMs are trained to distinguish a class against each of the other $k-1$ classes [49]. Class membership is decided upon which class is predicted by most classifiers.

2.1.2 Neural Networks

The SVM is defined by a subset of the training data samples $\mathbf{x}_i$ together with fixed basis functions like the RBF kernel (2.6). In the case of Neural Networks (NNs) the number of basis functions is a parameter of model selection, and the basis functions themselves are trainable. Much like their biological counterpart, NNs consist of uniformly behaving artificial neurons which are organized in layers. While there is a multitude of different possible topologies when building neural networks this chapter will only cover *feed-forward* neural networks without cycles. We will further focus on Convolutional Neural Networks in 2D as well as 3D [58].

Perceptron

The foundation for the concept of neural networks was laid 1958 by Frank Rosenblatt with the invention of the perceptron algorithm [82]. Rosenblatt introduced the concept of transforming a given real-valued input $\mathbf{x}$ using a nonlinear threshold function to define a two-class model of the form

$$z(\mathbf{x}) = f(\mathbf{w}^T \mathbf{x} + b)$$

with the so-called *weight vector* $\mathbf{w}$, *bias* b and the threshold function

$$f(\cdot) = \left\{ \begin{array}{ll} +1, & \cdot \geq 0 \\ -1, & \cdot < 0. \end{array} \right.$$

The two classes are represented by target values $+1$ and -1. We define $x_{j,i}$ to be the jth value of the ith training set sample with $j = 1, \ldots, J$ where J is the number of values of a sample. By defining $x_{0,i} = 1$ the weight w_0 will become the bias term b. Further the perceptron criterion is defined by

$$E_P(\mathbf{w}) = -\sum_{i \in \mathcal{M}} \mathbf{w}^T \mathbf{x}_i y_i,$$

where $\mathcal{M}$ are the indices of all samples $\mathbf{x}_i$ that are misclassified by the current model. To find weights $\mathbf{w}$ that separate the two classes weight updates can be computed with the update rule

$$\mathbf{w}^{t+1} = \mathbf{w}^t - \eta \nabla E_P(\mathbf{w}),$$

where η is called the *learning rate* and t is an index over the algorithm time steps. If two classes are linearly separable the perceptron algorithm is guaranteed to find the solution to the classification problem in a finite number of steps. Figure 2.2 shows the common schematic representation of a perceptron model.

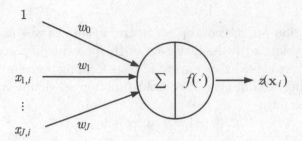

Figure 2.2: Schematic representation of the perceptron model. This
illustration with weights leading into a node that sums
and transforms the input is common for representing the
perceptron as well as neural networks.

Feed-forward Neural Networks

A feed-forward neural network can be imagined as a set of interconnected
neurons that perform a functional transform of N input variables $\mathbf{x}_i$

$$a_j^l = \sum_{i=1}^{N} w_{ij}^l \mathbf{x}_i + b_j^l. \tag{2.8}$$

where l is the layer index, the so-called *activations* a_j, $j = 1, ..., M$,
are the result of the linear combination of weights w_{ij}^l and the input
$\mathbf{x}_i$ together with a bias term b_j^l. For successive layers (2.8) becomes

$$a_j^l = \sum_{i=1}^{N} w_{ij}^l a_i^{l-1} + b_j^l,$$

with $a_i^0 = \mathbf{x}_i$, where the activations now are linear combinations of
activations of the previous layer. For a network with one *hidden* layer
like the one depicted in Figure 2.3, a complete linear feed-forward NN
can be defined as

$$z_j(\mathbf{x}, \mathbf{w}, \mathbf{b}) = \sum_{i=1}^{N} w_{ij}^2 \left(\sum_{k=1}^{M} w_{ki}^1 \mathbf{x}_k + b_i^1 \right) + b_j^2 \tag{2.9}$$

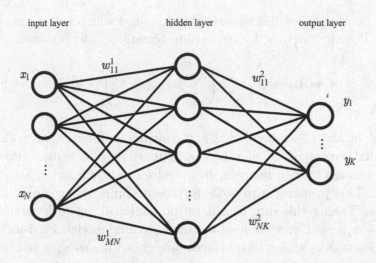

input layer   hidden layer   output layer

Figure 2.3: Example of a feed-forward neural network with three layers. Circles depict the neurons of different layers connected by weights. The biases b_i^l are not pictured.

given some input $\mathbf{x}$, network weights $\mathbf{w}$, and network biases $\mathbf{b}$, for $j = 1, ..., K$ outputs. Layers like in the ones in Figure 2.3 are called fully-connected because each neuron of the preceding layer is connected to each neuron of the subsequent layer.

So far the function transforms from layer to layer are linear. Since multiple linear transforms may be replaced by a single linear transform the hidden layer from (2.9) is dispensable. To be able to model functions that vary non-linearly with respect to the input non-linear activation functions are introduced. Using

$$\sigma(\cdot) = \frac{1}{1 - e^{-\cdot}}, \qquad (2.10)$$

which is often called the *sigmoid* function, and which is a special case
of the logistic function, for activation, equation (2.9) becomes

$$z_j(\mathbf{x}, \mathbf{w}, \mathbf{b}) = \sum_{i=1}^{N} w_{ij}^2 \sigma \left(\sum_{k=1}^{M} w_{ki}^1 x_k + b_i^1 \right) + b_j^2. \tag{2.11}$$

Hornik et al. [44] showed, that any continuous function on a closed
bounded subset of $\mathbb{R}^n$ may be approximated by a neural network,
given enough hidden neurons, to a desired approximation accuracy
$\varepsilon > 0$. This property also holds for the mapping of arbitrary finite
spaces. Taking this into consideration a neural network should be
able to represent every desired machine learning model. Feed-forward
neural networks with multiple layers are called *multi-layer perceptron*
or MLP.

Gradient Descent Optimization

Training a neural network means adapting weights as well as biases
so that the network *loss function* (which we will introduce in Section
2.1.2) is minimized. For now let us consider some loss function
$\mathcal{L}(\mathbf{w}, \mathbf{b})$, for which we try to find $\mathbf{w}$ and $\mathbf{b}$ so that the loss becomes
small. By computing the gradient of $\mathcal{L}(\mathbf{w}, \mathbf{b})$ with respect to $\mathbf{w}$ and
$\mathbf{b}$ one can update the parameters $\mathbf{w}$ and $\mathbf{b}$ in the opposite direction
of the gradient $-\nabla \mathcal{L}(\mathbf{w}, \mathbf{b})$ to minimize the loss. The learning rate
η determines the size of the steps one takes opposing the gradient.
Simple gradient descent, in this case for weights $\mathbf{w}$, can thus be defined
as an iterative process

$$\mathbf{w}^{t+1} = \mathbf{w}^t - \eta \nabla \mathcal{L}(\mathbf{w}^t) \tag{2.12}$$

where $\nabla \mathcal{L}(\mathbf{w}^t)$ is the gradient of the loss function with respect to $\mathbf{w}$
at time step t. Weights $\mathbf{w}^t$ get updated by small steps in negative
gradient direction and result in new weights $\mathbf{w}^{t+1}$. Biases $\mathbf{b}$ are
updated similarly by computing the derivative of the loss function
with respect to $\mathbf{b}$. Like for the perceptron algorithm η is the learning

rate. One iteration of gradient descent for the complete training data set is called one *epoch*.

The relation between loss and weights/biases is nonlinear. The graph of the loss function is a geometric surface, which might have many sinks, or local minima, as well as saddle points, or local maxima and plateaus. The goal of network training now is to manoeuvre on the loss surface finding a deep local minimum, while handling vanishing gradient. There exist a multitude of algorithms for gradient descent optimization of which we will only cover *stochastic gradient descent* and the *Adam* optimizer [48]. Neural network loss usually is computed either over single examples of the training set or so-called *mini-batches* which include only a subset of the training data. Stochastic gradient descent optimization is equivalent to (2.12) while computing gradients using single samples or mini-batches [1]. Usually the learning rate η will be adapted over time to get smaller for higher epochs. To be able to manoeuvre the loss function, gradient steps have to be chosen small. Small steps make stochastic gradient descent a slow optimization technique. By introducing *momentum*, past gradient information is incorporated into the gradient update step. For Adam optimization, the relation of first-order and second-order moments are used to adapt the current gradient update step.

Backpropagation

For training, the loss function gradients for each neural network parameter need to be computed. While information flow to this point has been forward, from the input layer to the loss function, *backpropagation* inverts this information flow to compute gradients. For backpropagation, the partial derivative of the loss function is passed through the network up to the neural network parameter for the update. We exclude activation functions and biases to simplify

---

[1] While one might suspect a random component due to the word "stochastic" this is not always the case.

the explanation of the backpropagation model. Consider a linear combination

$$a_j^l = \sum_i w_{ij}^l a_i^{l-1}$$

of weights w_{ij}^l and input a_i^{l-1} like for the neural network model in Fig. 2.3. We further define the loss at the output of a NN with L layers to be

$$\mathcal{L} = \sum_j (a_j^L - y_j)^2,$$

with $a_j^L = z_j(\mathbf{x}, \mathbf{w})$, where y_i either represents a class or regression value. We can thus compute the error signal at the output as the partial derivative of the loss with respect to a_j^L

$$\delta_j^L = \frac{\partial \mathcal{L}_n}{\partial a_j^L},$$

where $\mathcal{L}_n$ is the loss given sample n from the training data set. By using the chain rule of derivatives on δ_j together with the derivative of the activation with respect to the weights

$$k_i^L = \frac{\partial a_j^L}{\partial w_{ij}^L},$$

we can write the partial derivative of $\mathcal{L}_n$ with respect to the weights w_{ij}^L as a product of partial derivatives:

$$\frac{\partial \mathcal{L}_n}{\partial w_{ij}^L} = \frac{\partial \mathcal{L}_n}{\partial a_j^L} \frac{\partial a_j^L}{\partial w_{ij}^L}.$$

If we consider the sum over $w_{ij}^L a_i^{L-1}$ as neuron input and a_j^L as neuron output we are now able to describe error propagation though neurons by the multiplication $\delta_j^L k_i^L$. To compute the derivatives for early hidden layers (i.e. small l), we need to collect every δ_j^l of later neurons

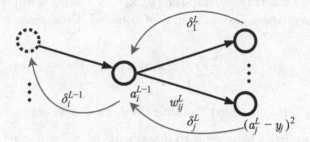

Figure 2.4: Outline of the propagation of errors δ backward through the NN. Red arrows indicate the direction of error propagation while the black arrow indicates the direction of information in forward propagation.

that have a link to the current neuron. By again applying the chain rule we arrive at

$$\delta_i^{l-1} = \sum_j \frac{\partial \mathcal{L}}{\partial a_j^l} \frac{\partial a_j^l}{\partial a_i^{l-1}}. \tag{2.13}$$

From (2.13) we can see that backpropagation enables derivative evaluation at arbitrary nodes in the network. This way one can now compute parameter updates for gradient de-scent for any network parameters. For general application each functional element (e.g. activation function) of the neural network needs to be differentiable. There do exist elements that are not differentiable like the max-pooling layer which will be introduced later in this Section. This problem is solved by defining element-specific rules on how to handle the gradient at the element in backpropagation.

Regression, Bernoulli and Multinoulli Neural Network Output

For regression, the neural network represents a function of the form $f : \mathbb{R}^n \to \mathbb{R}$. Regarding the input and output, the neural network needs to predict a continuous value z for some input $\mathbf{x}$ given experience inferred from the training dataset $(y_i, \mathbf{x}_i)$. A common approach is to define the regression loss as a sum of squared distances

$$\mathcal{L} = \frac{1}{N} \sum_{i=1}^{N} (\mathbf{z}(\mathbf{x}, \mathbf{w}, \mathbf{b}) - y_i)^2. \qquad (2.14)$$

Minimizing (2.14) is called training the neural network.

Transitioning from regression to classification, we will start with the prediction of a binary variable, or Bernoulli distribution, meaning that $y = 0$ stands for belonging to one class and $y = 1$ for belonging to the other class. In the case of this Bernoulli distribution, the neural network needs to output a probability from the interval $[0, 1]$. By applying the sigmoid activation from (2.10) the output is "squashed" to $[0, 1]$. While using the sigmoid together with (2.14) enables us to define a valid loss function for binary classification, the sigmoid's saturation for high and low values might restrain gradient-based learning. By introducing the *cross-entropy loss*,

$$\mathcal{L} = - \sum_{i=1}^{N} \{y_i \log(a_i) + (1 - y_i) \log(1 - a_i)\}, \qquad (2.15)$$

this saturation is countered and the magnitude of the loss gradient with respect to neural network weights is directly related to output error.

In case of an K-class classification problem one can use the *softmax* function to create network outputs that represent probability distributions over k classes also called multinoulli distribution. The softmax is defined as

$$\text{softmax}(\mathbf{z}_j) = \frac{e^{z_j}}{\sum_{i=1}^{K} e^{z_i}}.$$

Like for Bernoulli output, the cross-entropy loss can be applied to the multinoulli softmax output to stabilize gradient-based learning.

Convolution and Pooling

Given an input representation of grid-like structure like the two-dimensional image space or three-dimensional voxel space (see Section 4.2.2) natural elements or scenes in these spaces usually have a robust local correlation in a certain pixel-/voxel-neighbourhood. While in fully-connected layers each neuron of a preceding input layer interacts with each neuron of a successive layer, we will now define interactions with filter kernels. Filter kernels organize weights in small kernels that are trained to consider the local structure in some neighbourhood. Let us start by defining *convolution* in the context of neural networks. Convolution of two functions is defined as

$$(f * g)(x) = \int_{-\infty}^{\infty} f(\tau)g(x - \tau)d\tau.$$

This convolution can be imagined as the intregral of the product of $f(\tau)$ and a flipped version of funcion $g(\tau)$. The result is the $g(\tau)$-weighted average of $f(\tau)$. In case of neural networks the function $f(\tau)$ is the input image that gets convolved with the filter kernel $g(\tau)$. Discretized and extended to two-dimensional space convolution of an input image $\mathbf{I}$ with a two-dimensional filter kernel $\mathbf{K}$ is defined as

$$(\mathbf{I} * \mathbf{K})(i, j) = \sum_m \sum_n \mathbf{I}(m, n)\mathbf{K}(i - m, j - n). \qquad (2.16)$$

In practice convolution is implemented to be the cross-correlation function

$$(\mathbf{I} * \mathbf{K})(i, j) = \sum_m \sum_n \mathbf{I}(i + m, j + n)\mathbf{K}(m, n).$$

By sliding the filter kernel along each dimension of the 2d/3d input the same kernel is applied at each input pixel/voxel. This is why convolutional layers are regarded to have shared weights. The increment on the filter kernel position is called *stride*. Each filter kernel holds its own bias term. Sharing weights drastically reduces the amount of storage capacity required, and leads to a better generalization

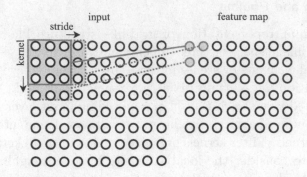

Figure 2.5: Convolutions of a filter kernel (blue) with the input. Each
convolution of the filter kernel with the input results in a
feature map entry. Dotted frames depict different locations
of the filter kernel after a stride step.

when compared to fully connected layers. After some input has
been convolved with a filter kernel and passed through the activation
function the result is called a *feature map*. Every new filter kernel
location yields an entry in the feature map. Figure 2.5 displays this
process for one filter kernel at different locations. Depending on the
definition of convolution at boundaries the resulting feature maps are
the same size as the input or slightly smaller.

Classical convolutional layers do not only consist of filter kernels but
also include *pooling* operations. Pooling can be seen as a form of
feature map compression, where a pooling operation decreases the
feature map size while retaining most of the information. We will only
cover the so-called *max pooling* operation since this is the only pooling
operation used in this thesis. Let us define the 2d pixel neighbourhood
$q \times q$ at some index pair i, j containing all pixels in the index pair
interval $[i, i + q - 1], [j, j + q - 1]$. A max pooling operation in a pixel
neighbourhood around an index pair i, j is defined as

$$p = \max \mathbf{I}(i, j).$$

Like convolution the pooling operation is applied at different indices depending on the stride. Choosing e.g. $q = 2$ and a stride of 2 a feature map is reduced to a quarter of its original size. The typical order of operations in a convolutional layer is to first apply filter kernels, then introduce a nonlinear activation function and apply pooling to the result.

The term *convolutional neural network* (CNN) is typically used for neural networks that consist of a set of convolutional layers followed by a set of fully connected layers. A well-known example of a CNN for image classification is the VGG19 network model by Simonyan et al. [92] (see Figure 2.8. To be able to connect convolutional layers and fully-connected layers the filter maps must be flattened. The flatten operation unrolls a matrix of arbitrary dimension to become a one-dimensional vector. Most of the times matrix and vector elements of the neural network's computational pipeline are simply called *tensors*. We will consider tensors to be numbers arranged in regular grids with arbitrary dimensionality. In the example in Figure 2.6 a two-dimensional input tensor is processed. While, e.g. grayscale images are two-dimensional tensors, RGB color images are three-dimensional tensors. In the case of the three-dimensional tensor, the two-dimensional convolution output will be summed over the third dimension to yield a two-dimensional feature map. The same rule applies for successive convolutional layers where the number of feature maps represents the third dimension. While applied to three-dimensional tensors this is still regarded as two-dimensional convolution since dimensionality is coupled to the convolutional kernel. If this kernel becomes three-dimensional, we will speak of a three-dimensional convolution layer. For three-dimensional convolution, one would sum over the fourth tensor dimension.

Regularization

The term *generalization* describes how well a specific machine learner can adapt to new data which has not been part of the training dataset.

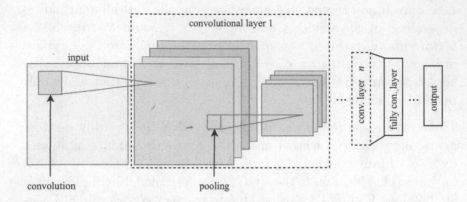

Figure 2.6: Schematic overview of a convolutional neural network.
Convolutional layer 1 consists of four filter kernels resulting
in four feature maps. Each feature map is pooled and
becomes the input to successive layers. After n consecutive
convolutional layers the output is flattened and passed
to a set of fully connected layers. The number and size
of filter kernels usually varies between layers. The size of
the flattened output at convolutional layer n is the number
of feature maps times their respective size. Intermediate
activations are not pictured.

Achieving high performance on the training data set while performing
poor on previously unseen test data is known as *overfitting*. The
prediction error of a model over independent previously unseen test
samples is often referred to as the generalization error. Techniques
that support generalization while preventing overfitting are known as
methods for *regularization*. While there exists a large body of work
on regularization we will only cover *dropout*, *early stopping*, as well as
batch-normalization.

Initially proposed in [93] by Srivastava et al., dropout is a technique
where random neurons are dropped from the neural network graph.
They temporarily lose all their connections and are ignored in the
forward pass as well as for backpropagation. At each epoch, a certain

percentage of neurons are randomly selected to be dropped. This counters the problem of co-adapting neurons in which the response of groups of neurons becomes coupled, and neurons start relying on the output of particular other neurons. Dropout can also be interpreted as training an ensemble of different networks which inherit a randomly selected group of weights from the network of the previous epoch.

Deciding on the number of epochs a neural network will be trained for is a challenging task. The definition of a stop criterion for training.is also called early stopping. If the the number of epochs is chosen too small the model may perform poor since it is not able to sufficiently adapt to the problem, which is also called underfitting. If the neural network is trained for too many epochs it may overfit. A simple way to define a stop criterion is to stop the training process if the generalization error gets larger instead of smaller for a certain number of epochs [79].

As for optimization by gradient descent, the learning rate, as well as initial parameters of the NN, need to be selected carefully. When propagating an error signal through the network, the input to each layer depends on each later layer. Changes in input distribution lead to the so-called covariance shift. Elimination of this covariance shift leads to faster training. *Batch normalization* introduced by Ioffe and Szegedy [46] uses normalization in-between layers to adjust input variance and mean. This way gradients become decoupled from parameter scale, and this allows for higher learning rates. Each scalar feature of a layer is normalized to have unit variance and zero mean. Each mini-batch leads to a separate set of estimates for feature variance and mean. Additionally, parameters that shift and scale the normalized features are introduced. These are learnable parameters which help keep representational power of the network.

2.1.3 Neural Texture Synthesis

A 2d Texture may be defined as a visual pattern modeled by a stationary stochastic process in image space. The goal of *texture synthesis* is to generate a synthetic new texture that aesthetically appears to have been generated using the same process that generated the sample. The resulting synthetic texture should be similar to the input in term of stochastic and structural composition. The quality of synthetic texture is usually evaluated visually by a human observer. See Figure 2.7 for a texture synthesis example.

sample texture   synthetic texture

Figure 2.7: Neural texture synthesis example from the supplementary material of [31].

Recently deep neural networks have been proposed for texture synthesis. A neural network is called "deep" if it consists of many layers [2]. In this section we will give an overview of the so-called *neural texture synthesis* approach introduced by Gatys et al. [31].

The main idea of neural texture synthesis is to derive image statistics of the different layers of a neural network and optimize an image to satisfy these statistics using the neural network. The neural network used in [31] is the VGG19 network introduced in [92] and shown in Figure 2.8. Based on a spacial summary statistic computed using the

---

[2]The term deep neural network is used loosely since there is no consensus on how many layers are needed for a deep neural network.

Figure 2.8: The VGG19 convolutional neural network. Each convolutional layer (grey) is shown with its name and the number of feature maps. The filter kernel size 3 × 3 is the same for each layer. The pooling layers (orange) use maximum pooling with non-overlapping 2 × 2 regions. The linear layers in blue are irrelevant for texture synthesis and thus not further detailed.

feature maps of different layers of VGG19 a new texture is synthesized that satisfies these statistics.

Following we will give an overview of the different steps of neural texture synthesis:

1. Activate an instance of the VGG19 network, called the reference network, with some texture x for synthesis.

2. Compute correlations of the feature maps within distinct layers. As we will show in Chapter 5 only a few layers are needed for a successful texture synthesis.

3. Activate a second instance of the VGG19 network, called the synthesis network, with an initialization image $\hat{x}$.

4. Optimize the initialization image using the synthesis network applying gradient descent to match the feature map correlations of the reference network.

A schematic overview of the setup for texture synthesis is given in Figure 2.9.

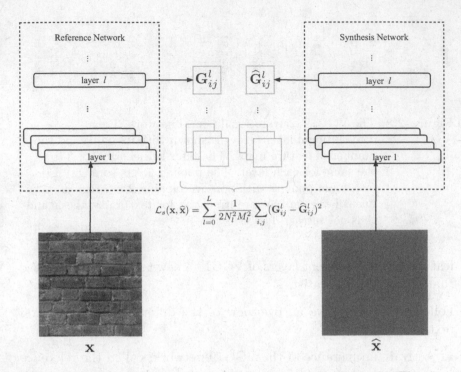

Figure 2.9: Schematic overview of the texture synthesis setup. The
reference network on the left provides the reference
Gramians (orange boxes). The synthesis network on the
right optimizes the input image $\hat{x}$ so that the loss $\mathcal{L}_s$
becomes minimal.

The feature map correlations of step 2 are given by the Gram-matrix
or Gramian

$$\mathbf{G}_{ij}^l = \sum_k \mathbf{F}_{ik}^l \mathbf{F}_{jk}^l, \tag{2.17}$$

where $\mathbf{F}_{ik}^l$ and $\mathbf{F}_{jk}^l$ are the flattened feature maps i and j of the layer
l, and k is the index of elements of the resulting feature map vector.
These inner products of filter activations of different layers are then

used to define a synthesis loss

$$\mathcal{L}_s(\mathbf{x}, \widehat{\mathbf{x}}) = \sum_{l=0}^{L} \frac{1}{2N_l^2 M_l^2} \sum_{i,j} (\mathbf{G}_{ij}^l - \widehat{\mathbf{G}}_{ij}^l)^2. \tag{2.18}$$

Since this can been seen as an averaging over feature map positions global texture coherence is lost. Using standard back-propagation, $\widehat{x}$ is optimized so that (2.18) becomes minimal, which leads to the synthetic texture $\widehat{x}$.

2.2 Parametric Curves and Surfaces

Let us define a parametric curve as a function

$$C : t \rightarrow (x(t), y(t)),$$

which maps a real parameter t from the interval $[0, 1]$ to $\mathbb{R}^2$ using the polynomials $x(t), y(t)$. While t can in fact be from any closed interval, $[0, 1]$ has become a common consensus. In case of parametric surfaces we introduce a second parameter s and define

$$S : (s, t) \rightarrow (x(s, t), y(s, t), z(s, t)),$$

which maps the real parameter pair (s, t) to $\mathbb{R}^3$. The parameter t and the parameter pair s, t map from 1d/2d domain to 2d/3d affine space of the curve or surface While polynomials are a good choice due to properties like differentiability, ability to adapt to a wide range of shapes as well as the existence of a multitude of algorithms, e.g. interpolation or approximation, using them to describe curves and surfaces also has drawbacks. Complex shapes require polynomials of high degree which are hard to control and lead to wiggly interpolations and approximations. Using the concept above, it is also un-intuitive to design polynomials that resemble a particular shape.

Many early works in the field of geometric curve and surface design have their roots in French car manufacturing. In 1959 Paul de

Casteljau working at Citröen started working on polynomial curves. Shortly after de Casteljau, Pierre Bézier at Renault also started working on these curves that became known as *Bézier curves*. While slightly ahead of time, de Casteljau's works [18, 19] on the topic were only discovered as late as 1975.

In the following sections, we will introduce the concepts developed by de Casteljau and Bézier whose definition is motivated by their application in geometric modeling. An introduction to the algorithm of de Casteljau is given in Section 2.2.1 followed by Bézier curves is Section 2.2.2. We give an introduction to B-spline curves and surfaces in Sections 2.2.3 and 2.2.4 and explain the concept of T-spline surfaces in Section 2.2.5.

2.2.1 The algorithm of de Casteljau

In compliance with G. Farin on the topic [26], we start by introducing the definition of a parabola using linear interpolation. Given a set of three points c_0, c_1, c_2 from $\mathbb{R}^2$ and a scalar parameter u we define

$$c_0^1(u) = (1 - u)c_0 + uc_1,$$
$$c_1^1(u) = (1 - u)c_1 + uc_2,$$
$$c_2^0(u) = (1 - u)c_0^1(u) + uc_1^1(u).$$

Aggregating this in a single formula we arrive at

$$c_2^0(u) = (1 - u)^2 c_0 + 2u(1 - u)c_1 + u^2 c_2. \qquad (2.19).$$

As can be seen from (2.19) we have arrived at a function quadratic in u defining a parabola by using linear interpolations. For u in the interval $[0, 1]$, the curve $c_2^0(u)$ will be fully contained in the convex hull defined by c_0, c_1, c_2. These so-called *control points* determine the shape of the curve. The polyline through the control points is called the *control polygon*. This concept can be generalized to curves of arbitrary degree n:

$$c_i^r(u) = (1 - u)c_i^{r-1}(u) + uc_{i+1}^{r-1}(u) \left\{ \begin{array}{l} r = 1, ..., n \\ i = 0, ..., n - r \end{array} \right. \qquad (2.20)$$

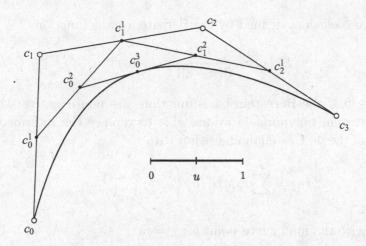

Figure 2.10: An example of a cubic Bézier curve evaluated at $u = 0.5$. The de Casteljau algorithm results in the final curve point c_0^3.

with parameter $t \in \mathbb{R}$, $c_i^0(u) = c_i$ and control points $c_0, c_1, ..., c_n \in \mathbb{R}^3$. As a final result of the linear interpolation, the parameter u results in curve point $c_0^n(u)$. While the algorithm itself is called the *de Casteljau algorithm*, the resulting curve is called a *Bézier curve*. Since the de Casteljau algorithm is based upon affine invariant operations, Bézier curves are invariant under affine transformations. Another important property of Bézier curves is the endpoint interpolation with $c^n(0) = c_0$ and $c^n(1) = c_n$. See Figure 2.10 for an example of a Bézier curve.

2.2.2 Bezier Curves

Given a set of control points $c_0, c_1, ..., c_n$ we define the Bernstein form of a Bézier curve as

$$C(u) = \sum_{i=0}^{n} c_i B_i^n(u),$$

with the coefficient defined by the Bernstein basis functions

$$B_i^n(u) = \frac{n!}{i!(n-i)!} u^i (1-u)^{n-1}.$$

For $u \in [0, 1]$ the Bernstein basis functions are non-negative. Using the Bernstein polynomials we are able to replace the intermediate points of the de Casteljau algorithm with

$$c_i^r(u) = \sum_{j=0}^{r} c_{i+j} B_j^r(u) \left\{ \begin{array}{l} r = 0, ..., n \\ i = 0, ..., n-r \end{array} \right. ,$$

and receive the final curve point for $r = n$.

Bézier curves have the following important properties:

Affine invariance: Affine transformations of a Bézier curve may be constructed by the same transformation of the control points.

Variation diminishing property: When intersecting a planar control polygon with a straight line the number of intersections with the Bézier curve is the same or less.

Convex hull property: The Bézier curve is contained inside the convex hull of its control points.

Degree: A curve with $n+1$ control points is of degree n.

Partition of unity: For any $u \in [0, 1]$ the sum of the Bernstein basis functions is 1.

Endpoint interpolation: A Bézier curve will pass through c_0 as well as c_n.

2.2.3 B-spline Curves

Before proceeding with B-spline curves, we introduce the concept of *continuity*. Consider two curves $G(u)$ and $H(v)$ with $u \in [a, b]$ and $v \in [b, c]$ for which we want to define a smooth transition. We call these curves C^0 continuous if $G(b) = H(b)$. We further call them C^1 continuous if the first derivative at the joint is the same for both curves, C^2 continuous if the same holds for the second derivative, and C^n continuous if the first through the nth derivatives are continuous. If two curves are not C^k continuous, they will not be C^n continuous for $0 \leq k < n$. If two curves are C^k, continuous they also are C^n continuous for $0 \leq n < k$. This continuity is also called parametric continuity, because it requires smoothness of the curve as well as the smoothness of the parametrization.

When using Bézier curves, in CAD, the flexibility in design depends upon the degree of the curve. Additional control points enable the design of complex shapes but increase the curve degree. By using connected curves as segments to construct complex curves, the overall degree can be kept small. *B-spline* curves are a generalization of Bézier curves which enables us to define curves constructed of curve segments. Segments are the result of a subdivision of the parametric domain by so-called *knots* into knot intervals. Knot vectors may contain the same knot multiple times which is called knot multiplicity. If a knot is contained s-times in the knot vector it is of multiplicity s.

A B-spline curve of degree k is defined as

$$C(u) = \sum_{i=0}^{n} c_i N_i^k(u),$$

with a non-decreasing knot vector $\mathbf{u} = (u_0, \ldots, u_n)$, control points $c_i, \ldots, c_n$ and B-spline basis functions $N_i^k(u)$. For a so called *clamped* curve to start at the first control point and end in the last control point knots u_0 and u_n have multiplicity $s = k + 1$. While it is possible to assign a weight ≥ 0 and $\neq 1$ to each control point, by which the B-spline curve would become rational, only non-rational B-spline curves are considered in this work.

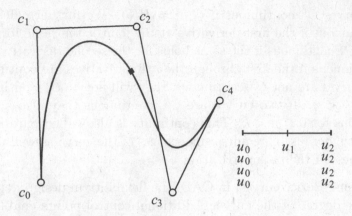

Figure 2.11: Example of a B-spline curve with $k = 3$ and knot vector $\mathbf{u} = (0, 0, 0, 0, 1, 2, 2, 2, 2)$. The solid square in the middle of the curve represents the joint of two curve segments.

When compared to Bézier curves the Bernstein basis functions have been replaced with the B-spline basis functions. The ith basis function of degree k is denoted $N_i^k(u)$ and defined by the *Cox-de Boor recursion*

$$N_i^0(u) = \begin{cases} 1 & \text{if } u_i \leq u < u_{i+1} \\ 0 & \text{otherwise,} \end{cases}$$

$$N_i^k(u) = \frac{u - u_i}{u_{i+k} - u_i} N_j^{k-1}(u) + \frac{u_{i+k+1} - u}{u_{i+k+1} - u_{i+1}} N_{i+1}^{k-1}(u), \tag{2.21}$$

with $u_i \in \mathbf{u}$. The subdivision of the parametric domain is achieved by making knots $\mathbf{u}$ a part of the basis function definition. The basis function $N_i^k(u)$ is non-zero on the interval $[u_i, u_{i+k+1}]$. By increasing the multiplicity of interior knots, the number of non-zero basis functions at this knot is reduced. Increasing knot multiplicity to s reduces continuity at the knot to C^{k-s}.

The following list contains some additional important properties of B-spline curves:

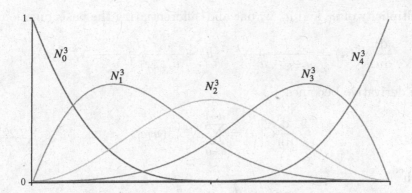

Figure 2.12: B-spline basis functions of degree 3 for the example in figure 2.11.

Strong convex hull property: The B-spline curve is completely contained inside the convex hull of its control points. Additionally the point $C(u)$ with $u \in [u_i, u_{i+1}]$ is completely contained in the convex hull of control points $c_i, ..., c_{i+k}$.

Local support: The shape of a B-spline curve segment is controlled by $k + 1$ control points. Changing any c_i will only affect curve points $C(u)$ with $u \in [u_i, ..., u_{i+k}]$.

Partition of unity: For any $u \in [u_0, ..., u_n]$ the sum of the B-spline basis functions is 1.

Since B-splines are a generalization of Bézier curves the Bézier properties from Section 2.2.2 also apply to B-splines. Figure 2.11 shows an example of a B-spline curve with the corresponding basis functions in Figure 2.12.

Derivatives of B-spline Curves

The derivative of a B-spline curve is again a B-spline curve of degree $k - 1$ with a set of new control points $c'_0, ..., c'_{n-1}$. Reducing the multiplicity of u_0 and u_n by one and differentiating the basis functions

$$\frac{\mathrm{d}}{\mathrm{d}u}N_i^k(u) = \frac{k}{u_{i+k} - u_i}N_i^{k-1}(u) - \frac{k}{u_{i+k+1} - u_{i+1}}N_{i+1}^{k-1}(u),$$

the derivative becomes

$$\frac{\mathrm{d}}{\mathrm{d}u}C(u) = \sum_{i=0}^{n-1} N_i^{k-1}(u)c_i', \qquad (2.22)$$

with

$$c_i' = \frac{k}{u_{i+p} - u_i}(c_{i+1} - c_i).$$

Higher order derivatives can be computed by iterating this concept.

We are also interested in the derivative of a B-spline curve with respect to a knot u_q like introduced in [74] by Piegl and Tiller

$$\frac{\mathrm{d}}{\mathrm{d}u_q}C(u).$$

The right derivative curve of a degree k B-spline curve is given by

$$C^r(u) = \sum_{i=0}^{n+1} \bar{N}_i^k(u)c_i^r, \qquad (2.23)$$

with

$$c_i^r = \begin{cases} 0, & i = 0, ..., q - k - 1 \\ \frac{1}{u_{i+p} - u_i}(c_{i-1} - c_i), & i = q - k, ..., q \\ 0, & i = q + 1, ..., n + 1. \end{cases}$$

The basis function $\bar{N}_i^k(u)$ is defined over a new knot vector $\bar{\mathbf{u}}$ which is the same as $\mathbf{u}$ but the multiplicity of u_q has been increased by one. As Piegl and Tiller state the left derivative $C^l(u)$ is identical to $C^r(u)$ as long as the knot u_q is of multiplicity one in $\mathbf{u}$. In this work, we assume that this is the case and that we are not interested in differentiating with respect to u_0 or u_n.

B-spline Curve Approximation

Suppose we have given a sequentially arranged set of points[3] $\mathbf{p} = (p_0, ..., p_m)$ for which we want to approximate a B-spline curve $C(u)$ of degree k with a non-decreasing knot vector $\mathbf{u} = (u_0, \ldots, u_n)$ where the knots u_0 and u_n have multiplicity $k + 1$ for end point interpolation. To compute the control points c_j of the B-spline curve $C(u)$ approximating $\mathbf{p}$, the least squares problem

$$\sum_{i=0}^{m} |p_i - C(t_i)|^2 \to \min \tag{2.24}$$

with parameters $\mathbf{t} = (t_0, \ldots, t_m)$ and end points $C(t_0) = c_0 = p_0$ and $C(t_m) = c_n = p_m$ is solved. To be able to solve (2.24) we need a parameter vector $\mathbf{t}$ and a knot vector $\mathbf{u}$. Computation of suitable $\mathbf{t}$ and $\mathbf{u}$ is the focus of Section 3.3. For now, we assume $\mathbf{t}$ and $\mathbf{u}$ are given. This yields the normal equation

$$(\mathbf{N}^T \mathbf{N})\mathbf{c} = \mathbf{q} \tag{2.25}$$

where $\mathbf{N}$ is the $(m-1) \times (n-1)$ matrix

$$\mathbf{N} = \begin{pmatrix} N_1^k(t_1) & \cdots & N_{n-1}^k(t_1) \\ \vdots & \ddots & \vdots \\ N_1^k(t_{m-1}) & \cdots & N_{n-1}^k(t_{m-1}) \end{pmatrix},$$

and $\mathbf{c}$ and $\mathbf{q}$ are the vectors defined as

$$\mathbf{c} = \begin{pmatrix} c_1 \\ \vdots \\ c_{n-1} \end{pmatrix}, \mathbf{q} = \begin{pmatrix} \sum_{i=1}^{m-1} N_1^k(t_i)q_i \\ \vdots \\ \sum_{i=1}^{m-1} N_{n-1}^k(t_i)q_i \end{pmatrix}$$

and

$$q_i = p_i - N_0^k(t_i)p_0 - N_n^k(t_i)p_m$$

---

[3]Sequentially arranged point clouds will be called point sets throughout this thesis.

for $i = 1, ..., m-1$. If there are no constraints for end point interpolation (2.25) reduces to

$$(\mathbf{N}^T\mathbf{N})\mathbf{c} = \mathbf{N}^T\mathbf{p}. \tag{2.26}$$

The control points c_j can be computed using (2.25), if

$$\sum_{l=1}^{m-1} N_i^k(t_l)N_j^k(t_l) \neq 0. \tag{2.27}$$

This is equivalent to the existence of a parameter $t_i \in [u_j, u_{j+1}]$ for $j = k, ..., n + 1$, see e.g. [26]. If the knot vector $\mathbf{u}$ includes all parametric values $\mathbf{t}$, the problem reduces to an interpolation of points $\mathbf{p}$.

2.2.4 Tensor Product B-spline Surfaces

Since B-spline curves are evaluated coordinate-wise, they are easily definable in $\mathbb{R}^3$. Here we can think of a surface as a curve swept through space while changing its shape. Surface definition can thus be achieved by moving the control points of one curve along a second curve. The resulting surface should possess the same properties as the underlying curve. Like Prautzsch et al. [78] we start by giving a universal definition of this concept. Given a set of parametric basis functions $G_0(u), ..., G_m(u)$ and $H_0(v), ..., H_n(v)$ we define a curve

$$C(u) = \sum_{i=0}^{m} g_i G_i(u),$$

for which the control points are defined by a second curve,

$$g_i = g_i(v) = \sum_{j=0}^{n} h_{ij}H_j(v)$$

which results in the surface definition

$$S(u, v) = \sum_{i=0}^{m}\sum_{j=0}^{n} g_i h_j G_i(u)H_j(v)$$

Since the basis functions are linearly independent their product results in a valid basis.

We are now able to define a *B-spline tensor product surface*

$$S(u,v) = \sum_{i=0}^{m} \sum_{j=0}^{n} c_{i,j} N_i^k(u) N_j^k(v), \qquad (2.28)$$

with individual knot vectors $\mathbf{u} = (u_0, \ldots, u_n)$, $\mathbf{v} = (v_0, \ldots, v_m)$, basis functions $N_i^k(u)$ and $N_j^k(v)$ as in (2.21) of degree k and $m+1$ rows of $n+1$ control points forming the control mesh. While it is possible to use basis functions with differing degree, we will use bicubic surfaces with $k = 3$ throughout this work. Start and end knots u_0, u_n and v_0, v_m have multiplicity $s = k+1$ which means that the surface corners will end in control points $c_{0,0}, c_{0,n}, c_{m,0}$ and $c_{m,n}$ (see figure 2.13 for a surface example). Almost all properties of B-spline curves also apply to B-spline tensor product surfaces except the variation diminishing property for which it is not clear how to even define the property in the context of surfaces.

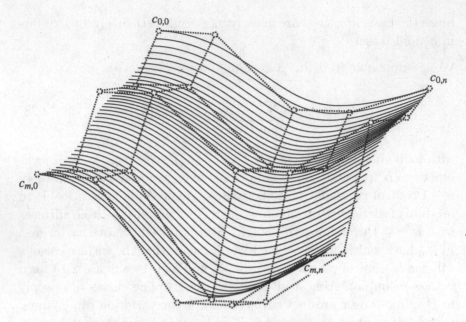

Figure 2.13: Example of a B-spline tensor product surface. The dotted
lines and points are the control mesh. Solid isolines
represent the surface. It is clear why this surface can be
imagined as a changing curve swept through space.

2.2.5 Tensor Product T-spline Surfaces

For B-spline surfaces[4] the control points have to lie on a perfect grid
where each row and column is continued across the control mesh. For
regions of high detail, where more control points are needed, this leads
to an increased overall number of control points even in regions where
fewer control points might suffice.

Introduced by Sederberg et al. [90] *t-spline surfaces* allow for so-called
T-junctions which can be seen as endpoints to rows and columns of
control points inside the control mesh. This way, control points may

---

[4]Since we exclusively examine tensor product surfaces we will hence use "surface"
instead of "tensor product surface"

be added in regions where they are required by design and not because of restrictions of the surface definition. While B-spline surfaces use global knot vector information, T-splines require knot vectors to be determined per control point. Knot vectors are computed by evaluating the pre-image of the control mesh. Knot information is based on so-called knot intervals which are assigned to edges in the rows and columns of the pre-image. Each edge of the pre-image represents a constant line in the direction of either u or v in the parametric domain. In Figure 2.14 the knot intervals of the pre-image

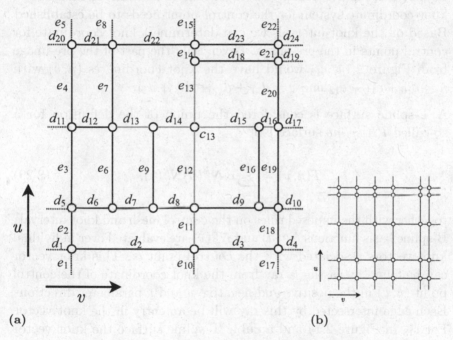

(a)   **(b)**

Figure 2.14: (a) Part of a pre-image in the u, v parameter domain of a T-spline surface with horizontal edges d_i and vertical edges e_i. Vertices in the pre-image correspond to control points of the control mesh. For clarity only control point c_{13} has been labeled. The relevant knot intervals for c_{13} in case of $k = 3$ are colored dark yellow. (b) Pre-image of a B-spline surface resulting from knot insertion into the grid from (a).

are denoted d_i and e_i. For them to be valid the knot intervals have to satisfy certain constraints:

- Knot intervals on opposing edges of a face need to have the same sum. In Figure 2.14 the equalities $e_7 = e_{13} + e_{14}$ or $d_{13} + d_{14} = d_{22}$ need to hold.

- If there exist opposing T-junctions with equal parametric value in either u or v they need to be connected.

To derive $\mathbf{u}_{i,j}$ and $\mathbf{v}_{i,j}$ knot vectors for individual control points a local knot coordinate system for the control point needs to be established. Based on the knot intervals we can determine a knot coordinate for control points in the pre-image. Given only the part of the pre-image from Figure 2.14, c_{13} would have the knot coordinates (u, v) with $u = e_{10} + e_{11} + e_{12}$ and $v = d_{11} + d_{12} + d_{13} + d_{14}$.

A T-spline surface is considered the union of the definition for a so-called *PB-spline* surface [90]

$$T(u,v) = \sum_{i=0}^{n} c_i N_i^k(u) N_i^k(v), \qquad (2.29)$$

together with the imposed rules on the control mesh and knot intervals. B-spline basis functions $N_i^k(u)$ and $N_i^k(v)$ are evaluated over individual knot vectors associated with the control point c_i. This knot vector can be found by casting a ray from the knot coordinate of the control point (u, v) in the positive and negative u and v parameter directions. Each edge intersected by this ray will be an entry in the knot vector. For c_{13} in Figure 2.14 and a cubic T-spline surface the knot vector becomes

$$\mathbf{u}_i = (u - e_{12} - e_{11}, u - e_{12}, u, u + e_{13}, u + e_{13} + e_{14})$$
$$\mathbf{v}_i = (v - d_{14} - d_{13}, v - d_{14}, v, v + d_{15}, v + d_{15} + d_{16}).$$

T-spline surfaces are a generalization of B-spline surfaces. For a rectangular grid without T-junctions, the T-spline surface becomes a B-spline surface. T-spline surfaces can be converted to B-spline surfaces by knot insertion so that no T-junctions remain. Figure 2.14b shows the pre-image result of such a conversion from T-spline to B-spline surface. T-spline surfaces are affine invariant, obey the convex hull property and, under certain conditions, satisfy the partition of unity property as well as the independence of basis functions [62].

3 Parametrization in Curve and Surface Approximation

For the approximation of B-spline or T-spline surfaces to point clouds the computation of a parametrization is required. Since the construction of both surface types is based upon the concept of B-spline curves, we begin by developing parametrization methods for curves. Parametrization for B-Spline curve approximation includes the computation of a point parametrization $\mathbf{t}$ as well as the knot vector $\mathbf{u}$ for a given set of ordered points. Since the point parametrization and the knot vector have a significant impact on the quality and tightness of the resulting approximation, the recovery of a suitable set of parameters is crucial in curve and surface approximation. A good parametrization leads to minimal deviation of the point set to the curve or surface. We define this deviation as the Hausdorff distance (HD) of the point set to the curve $h(\mathbf{p}, C(u))$ or the surface $h(\mathbf{p}, T(u,v))$. Additionally, the approximation should result in as few control points as possible which implies that the number of knots in $\mathbf{u}$ should stay small. Usually, a predefined error bound or threshold needs to be satisfied to guarantee a sufficient approximation quality. Most parametrization methods compute a point parametrization first and iteratively add knots to a previously empty knot vector until the threshold is satisfied in a second step. This approach to knot vector computation is called knot vector refinement.

In Section 3.2 an approach using SVMs to compute a knot vector is presented. The SVM is trained to assess locations along the point set for their suitability in becoming a split location. This split corresponds to adding the parametric value of the split location,

© Springer Fachmedien Wiesbaden GmbH, part of Springer Nature 2020
P. Laube, *Machine Learning Methods for Reverse Engineering of Defective Structured Surfaces*, Schriftenreihe der Institute für Systemdynamik (ISD) und optische Systeme (IOS), https://doi.org/10.1007/978-3-658-29017-7_3

or point, to the knot vector. Based on geometric features computed in the neighbourhood of a single point the SVM can assign a so-called score to the position. The topography of this score along the point set is used to compute the final knot vector. This score-based knot placement method (SKP) does not handle the computation of parametric values t.

In Section 3.3 a method for computing both, a point parametrization, as well as a knot vector using deep neural networks is introduced. Two interdependent neural networks of which one computes a suitable vector t and the other predicts new knot positions for knot vector refinement are presented. The neural networks directly operate on the point data without an intermediate feature vector. We will show that it is possible to include B-spline curve approximation directly into the neural network architecture. This way it is possible to define a loss for the neural networks based on approximation quality.

The machine learning approaches for B-spline curve approximation from Sections 3.2 and 3.3 are applied to T-spline surface approximation in Section 3.4. A discussion of the methods and their properties as well as their performance for B-spline curve and T-spline surface approximation is given in Section 3.5.

3.1 Related Works

3.1.1 Point Parametrization

Computing point parametrizations is usually based on three classical methods: *uniform*, *chord-length* and *centripetal* parametrization [59]. The uniform method spaces knots uniformly throughout the parametric domain. It does not incorporate the spacing of data points which leads to poor performance for points sets that are not uniformly spaced. The chord-length parametrization supposes that the distance between data points is close to the length of the approximated curve between two data points. The actual point parametrization is based upon the chord length ratio between two points and the sum of chord lengths of the point set. While the chord-length method works well for unevenly spaced point sets, it does not consider the fact that it is hard to parametrize polynomial curves to have unit speed [27]. This problem is addressed by the centripetal method. It extends the chord-length method by increasing the impact of smaller chords and decreases the impact of longer chords which leads to improved approximation quality on sharp turns. Throughout literature, the centripetal method is the most prominent of the three. A method closely related to the uniform method is the universal method introduced by Lim [63]. This methods is claimed to result in curves that appear more natural than classic parametrizations. Many other parametrization methods are combinations of newer approaches and classic parametrizations like the hybrid method proposed by Shamsuddin [91]. It combines the universal method with aspects of the centripetal method to produce smoother results.

3.1.2 Knot Placement

Methods for knot placement are either based on optimization, knot vector refinement or instant knot vector computation. Methods for instant knot vector computation instantly generate a knot vector given the desired number of knots. Refinement based methods include or remove knots from an initial knot vector until a specified error threshold is satisfied. Optimization-based methods are closely related to methods for refinement but optimize knot positions over time. Both refinement and optimization based methods are subject to iterations.

The best-known method for know placement is the averaging method by Piegl and Tiller [73]. They instantly compute a knot vector by averaging parametric values without using geometric features of the point set. The method leads to an even distribution of knots in the parametric domain and is strongly dependent on the point parametrization. Another method for instant computation using geometric information of the point set is the one by Razdan [81].

Razdan gives a detailed analysis of different geometric features for the selection of parametric values of parametrized points as knot values. Li et al. [61] instantly compute the knot vector using a heuristic. Knots are included where an angular threshold between adjacent points is violated. The refinement based method by Park and Lee [71] uses dominant points in the point set to select knot positions. They split the parametric domain by knots so that the resulting curve segments are of somewhat equal complexity. This complexity is measured by the shape index. There exists a large body of work about knot vector optimization. Methods using optimization range from non-linear optimization [12, 114, 87, 34] to genetic algorithms [112, 86, 98, 100, 99], elitist clonal selection [30] or meta-heuristics like the firefly algorithm by Gálvez et al [29]. Yuan et al. [113] use the Lasso method to extract optimal subsets from a multi-resolution B-spline basis. Kang et al. define knot placement as a convex optimization problem to reduce computational complexity.

3.2 SVMs for Knot Placement in B-spline Curve Approximation

In the following Sections, we describe our approach for knot vector computation using support vector machines. Our approach consists of three main stages:

1. The *training set generation* (Section 3.2.2).

2. The *feature extraction* of geometric features for points in a point set (Section 3.2.1).

3. The *knot selection* for computation of a knot vector using trained support vector machines (Section 3.2.3).

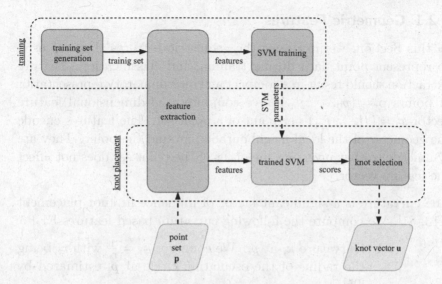

Figure 3.1: Overview of the different steps of the knot vector computation. General steps of the process are colored red, data entities are yellow, and the trained SVM is green. Point set **p** is the set for which a parametrization needs to be computed and is not part of the training data set.

Figure 3.1 gives an overview of our approach. First, a training set consisting of point sets and knot vectors is computed. For training and classification, a feature vector is computed. After the *SVM training* the SVM can assign a score to each point. While the SVM is trained for classification, only the score of each point will be used in succession. This score assesses each point location's impact on the approximation error if its corresponding parametric value is added as a knot to the knot vector. The topography of the score along the point set yields the final knot vector. After the training of the SVM no further optimization is required. Since no optimization is required during knot vector computation, this is a method for instant knot vector computation.

3.2.1 Geometric Features

In this Section, we introduce the geometric features that are used to represent point data during training and classification. Feature extraction should result in a compact and discriminatory representation of points $\mathbf{p} = (p_0, \ldots, p_m)$. We compute a 14-dimensional feature vector $x_i = (f_1^i, \ldots, f_{14}^i)$ per point of a point set. The features encode the geometry of the local neighbourhood around the point. They are symmetric so that inverting the order of the point set does not affect the feature vector.

The curvature is a significant factor of influence in knot placement [113, 81]. We compute the following curvature based features F1-F5:

F1: Curvature κ_i at p_i. We compute $\kappa_i = \frac{1}{r_i}$ with r_i being the radius of the osculating circle at p_i estimated by [95].

F2 & F3: Minimum and maximum approximate curvature derivatives to the left and right

$$f_2^i = \min\left(\frac{\nabla \kappa_i}{\nabla t_i}, \frac{\Delta \kappa_i}{\Delta t_i}\right), \quad f_3^i = \max\left(\frac{\nabla \kappa_i}{\nabla t_i}, \frac{\Delta \kappa_i}{\Delta t_i}\right),$$

with forward and backward difference operators Δ and ∇.

Since they have shown to be of great importance in knot placement [71] we use local curvature maximum (LCM) points to define LCM-based features. The set $\mathbf{p}^L = \{p_0^L, \ldots, p_s^L\} \subseteq \mathbf{p}$ of LCM points contains $p_0^L = p_0$ and $p_s^L = p_m$. The other LCM points $p_j^L = p_\ell$ are defined as

$$\kappa_\ell > \max(\kappa_{\ell-1}, \kappa_{\ell+1}) \quad \text{and} \quad \kappa_\ell \geq \kappa_{LCM}$$

with

$$\kappa_{LCM} = \tau \sum_{i=0}^{m} \kappa_i, \tag{3.1}$$

and

$$\tau = \frac{1}{4(m+1)}.$$

We further define an LCM segment as $\mathbf{p}^j = \{p_1^j, \ldots, p_{r_j}^j\}, j = 0, \ldots, s-1$ consisting of all r_j points p_i strictly between p_j^L and p_{j+1}^L, i.e.

$$\mathbf{p} = \mathbf{p}^L \,\dot\cup\, \mathbf{p}^0 \,\dot\cup \ldots \dot\cup\, \mathbf{p}^{s-1},$$

where $\dot\cup$ denotes the disjoint union. Thus, there is a one-to-one correspondence of points p_j^L and p_i^j to indices $\{0, \ldots, m\}$. We denote their parameters with $t(p_j^L)$ and $t(p_i^j)$, respectively. We compute the following LCM-based features:

> F4: Ratio of the approximate curvature integral over all points $\mathbf{p}$ and the approximate curvature integral over points $\mathbf{p}^j$ with $p_i \in \mathbf{p}^j$.

$$f_4^i = \frac{\sum_{\ell=0}^{r_j} \kappa_\ell.}{\sum_{\ell=0}^{m} \kappa_\ell}.$$

> F5: Mean curvature of the points $\mathbf{p}^j \cup \{p_j^L, p_{j+1}^L\}$.

Using angle-based heuristics in knot vector computation has positive effect on the approximation quality [61]. Based on this observation we include the following angle-based feature:

> F6: The angle $\sphericalangle(p_j^L, p_i^j, p_{j+1}^L)$.

Other important factors in knot placement are the chord length of a curve segment and parametric distances between knots. We compute the following distance-based features F7-F14:

> F7 & F8: Minimum and maximum of the Euclidean distances of point p_i^j to p_j^L and p_{j+1}^L.

> F9 & F10: Minimum and maximum of the approximate arc lengths between p_i^j and p_j^L as well as p_i^j and p_{j+1}^L as the sum of Euclidean distances between all intermediate points.

F11 & F12: Minimum and maximum of the parametric distances $\Delta_{ij} = t(p_j^L) - t(p_i^j)$ and $\nabla_{ij} = t(p_i^j) - t(p_{j+1}^L)$ relative to the parametric distance of p_j^L to p_{j+1}^L as

$$f_{11}^i = \min\left(\frac{\nabla_{ij}}{\Delta t(p_j^L)}, \frac{\Delta_{ij}}{\Delta t(p_j^L)}, \right),$$

$$f_{12}^i = \max\left(\frac{\nabla_{ij}}{\Delta t(p_j^L)}, \frac{\Delta_{ij}}{\Delta t(p_j^L)}, \right).$$

F13 & F14: Minimum and maximum of the parametric distances from p_{i-1}^j to p_i^j and p_i^j to p_{i+1}^j relative to the parametric distance of p_j^L to p_{j+1}^L

$$f_{13}^i = \min\left(\frac{\nabla t(p_i^j)}{\Delta t(p_j^L)}, \frac{\Delta t(p_i^j)}{\Delta t(p_j^L)} \right),$$

$$f_{14}^i = \max\left(\frac{\nabla t(p_i^j)}{\Delta t(p_j^L)}, \frac{\Delta t(p_i^j)}{\Delta t(p_j^L)} \right).$$

Each point set $\mathbf{p}$ results in $m - s$ feature vectors. Features are normalized to have unit-variance and zero-mean. The resulting design matrix would contain a row for each point p_i and 14 columns.

3.2.2 Training Set Generation

Finding a knot vector that contains only a few knots while resulting in tight approximations is a challenging task. Finding a good knot vector, e.g. by searching for knot combinations using exhaustive search in the parametric domain is computationally expensive due to combinatorial complexity.

We propose to use constrained exhaustive search to compute knot vectors for training. Our training set computation has the following steps:

1. Generate random control points c_i using the standard normal distribution with mean μ and variance σ to define clamped uniform B-spline curves of degree k with two interior knots. We shift the mean by $\Delta\mu > 0$ for x-coordinates of consecutive control points. Curves with self-intersections are discarded.[1] The uniform knot vector is only used to create the training set B-spline curves. The knot vector that will be used for actual training is created in step 3.

2. Sample $m + 1$ points $\mathbf{p} = (p_0, \ldots, p_m)$ uniformly along the curve and compute their centripetal parameters $\mathbf{t} = (t_0, \ldots, t_m)$ [59] [2].

3. Add parameter values t_0 and t_m of p_0 and p_m to the initially empty knot vector $\mathbf{u}$.

4. Compute the LCM points of $\mathbf{p}$ and add the corresponding parameter values to the *initial knot vector* $\mathbf{u} = (u_0, ..., u_w)$. [3]

5. Approximate $\mathbf{p}$ with a B-spline curve $\widetilde{C}$ of degree k and knot vector $\mathbf{u}$.

6. Insert further knots from $\mathbf{t}$ into $\mathbf{u}$ to minimize $h(\mathbf{p}, \widetilde{C})$.

7. Repeat Steps 5 to 7 until a predefined improvement rate ε is achieved.

8. Raise the multiplicity of the end-knots to $k + 1$.

---

[1] We discard self-intersecting curves to make sure that the sequential order of sampled points is unique. Self-intersections would lead to a split of the point set into subsets in reverse engineering.

[2] We also tested the chord length parametrization for the generation of the test data and knot placement. This lead to a decrease in the overall approximation quality.

[3] Overall our approach led to $1,656,000$ tested knot vectors over the training set with an average of 17 knots per curve. Searching for a combination of 17 knots out of 148 possible knot locations without the segmentation by LCMs would yield $8.48e21$ possible knot vectors to test. This would render the computation of a large enough training set infeasible.

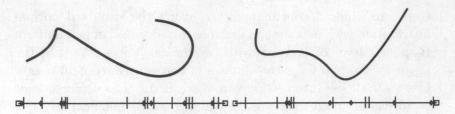

Figure 3.2: Two curve examples from the synthetic data set together
with their respective knot vectors. Diamonds are LCM knots
and vertical lines are knots found by exhaustive search.

We set parameters $k = 3$, $\mu = 1$ and $\sigma = 0.5$ for y-coordinates, $\mu = 1$
and $\sigma = 0.1$ for x-coordinates of the first control point and increment
μ by $\delta\mu = 0.1$ for consecutive control points. We further set $m = 149$
and $\varepsilon = 0.15$.

This process results in a diverse set of curves that contain sections
with little to no curvature as well as sections with high curvature.
Ideally, a data set consisting of diverse curves from real-world examples
would be used. At the time of writing no sufficiently large datasets
are publicly available. A set of $1,000$ training and 250 curves for
evaluation is computed. This dataset leads to a total of $173,889$ points
for training. While being used to compute feature vectors for points
p_i, LCM points are excluded from the data sets. Figure 3.2 shows
two examples from our training set.

3.2.3 Score-based Knot Placement

SMVs are trained to assign the following classes to individual points
based on their geometric features:

- Parametric value of the point has been selected to be a knot in
  exhaustive search (see Section 3.2.2).

- Parametric value of the point does not correspond to a knot value.

For now, we assume that model selection already resulted in the best-performing SVM. While this SVM could now be used to assign knot labels to points in the point set, there is no priority information on which knot to place first. Instead of using class labels directly to place knots we propose to use the score assigned to an input x_i by evaluating (2.7). While SVMs do not provide class membership probabilities, it has been shown that the score may be used to derive such probabilities [76, 101], which makes it a suitable measure for knot placement. The insertion of a knot in regions of high score along **p** will presumably be beneficial for approximation quality.

To compute a knot vector for a previously unseen point set, we start by computing LCM points $\mathbf{p}^L$. Then the feature vectors x_i for all $p_i \in \mathbf{p} \setminus \mathbf{p}^L$ are computed. After that, the score s_i for each x_i is evaluated, which results in the score vector **s**. Score is normalized to $[0, 1]$. For all $p_i \in \mathbf{p}^L$ we set $s_i = 1$. We also set $s_0 = s_m = 1$. Then score maxima in **s** are identified based on the topographic measures *relative maximality* and *prominence*. Relative maximality of a value s_i is given if $s_i > s_{i-1}, s_{i+1}$. Prominence measures the height of a score maximum s_i relative to surrounding local maxima. Denote by s_j and s_k the closest scores to the left and right of s_i with $s_j, s_k > s_i$. This means, $s_j, s_k > s_i$ for $j < i < k$ and all scores in the sub-vector $(s_{j+1}, \ldots, s_{k-1})$ are at most s_i. Then, the prominence r_i of s_i is defined as

$$r_i = s_i - \max\big(\min(s_j, \ldots, s_i), \min(s_i, \ldots, s_k)\big).$$

For knot placement the relative maxima with maximum prominence are selected. If this results in more than one score maximum the score maximum closest to the midpoint of the largest knot interval is selected. The parametric value t_j that corresponds to the selected score is added to the knot vector. This step is repeated until no relative maxima remain. Further knots are added based on the sum of scores. We split knot segments whose score sum is largest amongst all curve segments by introducing the parametric value at the median of the segment to the knot vector. This step is repeated until the desired number of knots is reached, or the error threshold is satisfied. This strategy is derived from orography: one can interpret the score as height in a mountainous region and pick peaks with the largest prominence first, before further subdividing whole regions based on height. This knot placement approach is laid out in detail in algorithm 1. This knot placement strategy ensures that every knot span contains at least one parametric value t_i. By that $\mathbf{N}^T\mathbf{N}$ from (2.26) is regular and that condition (2.27) is satisfied.

Noisy point clouds may lead to an incorrect identification and a larger number of LCM-points which increases the number of maxima in the score s and a decreasing influence of the SVM on the knot placement process. To counter an increasing number of LCM-points we propose to increase τ in (3.1) or to apply point cloud filters like [26] before LCM and feature computation for noisy point clouds.

Algorithm 1 Score based knot placement (SKP) for curves.

Require:
$\mathbf{t} = (t_0, ..., t_m)$, parameter vector
$\mathbf{s} = (s_0, ..., s_m)$, score vector
$\mathbf{r} = (r_0, ..., r_m)$, prominence vector
k, b-spline degree
c, target number of knots
$\#(\cdot)$, number of entries in a vector

function SKP($\mathbf{t}$, $\mathbf{s}$, $\mathbf{r}$, k, c)
    initialize $\mathbf{u}$ with 0 and 1
    $\mathbf{v}$:= vector of indices of relative maxima in $\mathbf{s}$
    while $c > 0$ **do**
        if $\#(\mathbf{v}) > 0$ **then**
            $\mathbf{v}_{\max}$:= $\underset{i \in \mathbf{v}}{\arg\max}(\mathbf{r})$
            if $\#(\mathbf{v}_{\max}) > 1$ **then**
                ℓ := $\underset{i=0,...,\#(u)-2}{\arg\max} (u_{i+1} - u_i)$
                $\bar{u}$:= $(u_\ell + u_{\ell+1})/2$
                j := $\underset{i \in \mathbf{v}_{\max}}{\arg\min}(|t_i - \bar{u}|)$
            else
                j := $(\mathbf{v}_{\max})_0$
            $\bar{t}$:= t_j
            $\mathbf{v}$:= $\mathbf{v} \setminus (j)$
        else
            ℓ := $\underset{i=0,...,\#(u)-2}{\arg\max} \left(\underset{t_j \in [u_i, u_{i+1}]}{\sum} s_j \right)$
            $\bar{u}$:= $(u_\ell + u_{\ell+1})/2$
            $\bar{t}$:= $\underset{t \in \mathbf{t}}{\arg\min}(|t - \bar{u}|)$
        $\mathbf{u}$:= $\mathbf{u} \cup (\bar{t})$
        $\mathbf{t}$:= $\mathbf{t} \setminus (\bar{t})$
        c := $c - 1$
    raise multiplicity of end-knots of $\mathbf{u}$ to $k + 1$
    return $\mathbf{u}$

3.2.4 Results

We apply a non-linear *Gaussian RBF kernel* function for the SVM. A
k-fold cross validation is used with $k = 5$. For optimizing the slack
variable C and kernel size γ, an extensive grid search is performed.
We use the following (C, γ)-set:

$$T = \{0.01, 0.1, 1, 100, 1000, 10000, 50000\}$$
$$\times \{0.01, 0.1, 0.25, 0.5, 1.0, 2, 5, 10\}.$$

Sampling the (C, γ)-space at exponentially growing values has become
a best practice for SVM grid search [45]. Each trained SVM is
evaluated by comparing HD at different numbers of knots $5, \ldots, 25$
using the method described in Section 3.2.3. A minimal mean HD
over the evaluation set is achieved for the SVM trained with $C = 0.01$
and $\gamma = 10$.

We compare our knot placement method with two well-known methods:
the knot placement method by Piegl and Tiller (NKTP) [73] and the
Dominant-Points-based method (DPKP) by Park et al. [71]. We
use the same curvature values without denoising for our Score-based
knot placement method (SKP) and the DPKP method. While NKTP
is a method for instant knot placement, the DPKP method uses
distance-based adaptive refinement and thus is not able to instantly
compute a knot vector. The methods are compared by average HD
over the validation set. Figure 3.3a shows the average HD over different
numbers of knots.

Approximation quality of the SKP method is superior to the NKTP
method. While results are very close to those of the DPKP method
the SKP method results in slightly higher HD over the test set. When
looking at the runtime of the methods in figure 3.3b one can see that
the SKP method has a significantly lower runtime, staying below
0.002 seconds, when compared to the DPKP method. The run time
of the DPKP method increases almost linearly with the number of
knots. Run-time of the SKP method excludes the training process
and SVM initialization since this has to be done only once beforehand.

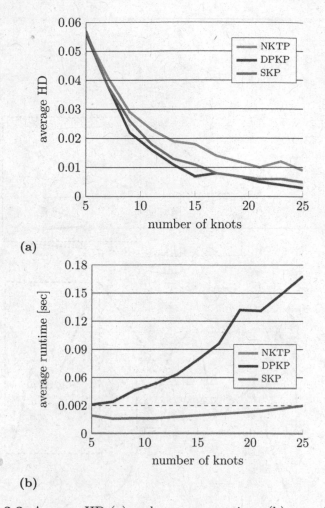

(a)

(b)

Figure 3.3: Average HD (a) and average run-times (b) over the test set at different numbers of knots for NKTP, DPKP, and SKP. The ordinate in b is of non-linear scale.

Figure 3.4 shows two examples of approximation by the different methods with 34 knots each. The SKP and DPKP method manage to follow the point set closely. The NKTP method results in low-quality

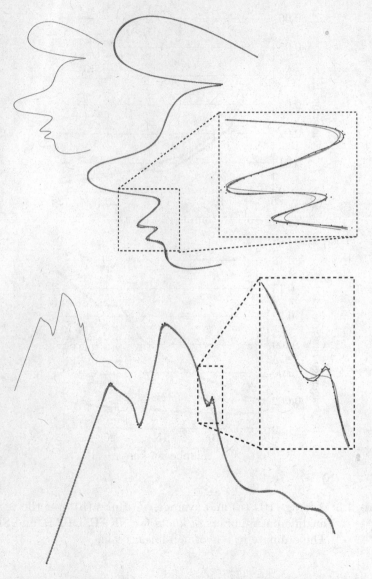

Figure 3.4: Approximations of two point sets by NKTP (orange), DPKP (blue), and SKP (red) with a close-up of the dashed region.

approximations especially for the complex regions from the close-ups. Remarks regarding the runtime can be found in appendix A.1.

3.3 Deep Learning Parametrization for Curve Approximation

In this Section, an approach to compute parametric values t and a knot vector u for B-Spline curve approximation using deep neural networks is presented. In order to be able to apply deep learning to the problem of parametrization, one has to address a number of requirements. Like in Section 3.2.2 we need to create a large enough training set that shares the characteristics of real data. Also, a suitable network loss for the parametrization of a B-spline curve needs to be defined. In comparison to Section 3.2 we will not use geometric features at single points of the point set as input but the coordinates of ordered two-dimensional point sets or point segments. Networks trained directly on point data [80] have shown exceptional performance in reconstruction and classification tasks. A feed-forward neural network, without convolutional layers, requires a fixed size input whereas point sets for approximation are of variable size. The approach needs to be able to cope with this problem.

We cover the parametrization approach in Section 3.3.1, the network architecture in Section 3.3.2, the training process in Section 3.3.3. Results are presented in Section 3.3.4.

3.3.1 Parametrization Approach

Figure 3.6 gives an overview of our parametrization approach. We follow Figure 3.6 in this exposition. Besides the pre- and post-processing steps our pipeline includes the following two interdependent NNs:

- Knot Selection Network (KSN)

- Point Parametrization Network (PPN)

These NNs are explained in detail in Section 3.3.2.

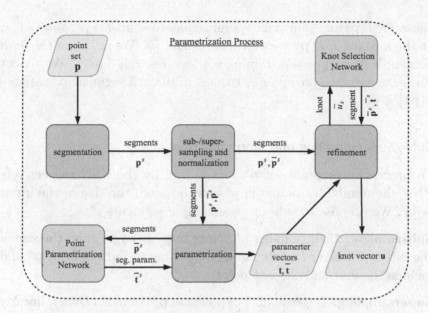

Figure 3.5: Overview of the parametrization process. Inputs and outputs are marked in yellow. Red boxes are part of pre-/postprocessing. The point parametrization and knot selection networks are explained in Section 3.3.2.

Segmentation

We define the complexity of a point set $\mathbf{p}$ to be the total curvature

$$\widehat{\kappa}(\mathbf{p}) = \sum_{i=0}^{m-1} \frac{(|\kappa_i| + |\kappa_{i+1}|)\|p_{i+1} - p_i\|_2}{2},$$

with curvature κ_i at point p_i. Like in Section 3.2.1 we compute curvature using osculating circles. Using $\widehat{\kappa}(\mathbf{p})$ we can define a measure for the training set complexity. Training set complexity serves as an estimate of the complexity the NN will be able to handle. Application of this complexity measure is specific to the presented approach. We propose to split point sets $\mathbf{p}$ into segments that match the complexity of the training dataset. For a point sequence $\mathbf{p}$ compute the total curvature $\widehat{c}(\mathbf{p})$ and split $\mathbf{p}$ into point sequence segments $\mathbf{p}^s$, $s = 1, ..., r$, at the median, if $\widehat{\kappa}(\mathbf{p}) > \widehat{\kappa}_t$ for a threshold $\widehat{\kappa}_t$. We set $\widehat{\kappa}_t$ to the 98th percentile of $\widehat{\kappa}(\cdot)$ of the training set (see Section 3.3.3). We repeat this segmentation process $r - 1$ times until each segment $\mathbf{p}^s$ satisfies $\widehat{\kappa}(\mathbf{p}^s) < \widehat{\kappa}_t$.

Sub-/Supersampling and Normalization

To process some point set $\mathbf{p}^s$, $s = 1, ..., r$, by the KSN respectively PNN the number of points in p^s has to match the size of the input layer. We propose to sub- or supersample segments p^s:

Subsampling: If the number of points in $\mathbf{p}^s$ is larger than the size of the NN input layer l, draw equally distributed indices from $\mathbf{p}^s$ and include the first and the last point.

Supersampling: If the number of points in $\mathbf{p}^s$ is smaller than l, linearly interpolate placeholder points between consecutive points p_i^s and p_{i+1}^s starting from $i = 0$. Iterate until the number of points is equal to l. Placeholder points are only used during knot placement and are discarded after knot vector computation.

We define

$$\bar{p}_i^s = \frac{p_i^s - \min(\mathbf{p}^s)}{\max(\mathbf{p}^s) - \min(\mathbf{p}^s)},$$

to be the normalized segments $\mathbf{p}^s$ after sub-/supersampling with $\min(\mathbf{p}^s)$ and $\max(\mathbf{p}^s)$ being the minimum and maximum coordinates of $\mathbf{p}^s$.

Parametrization

For each $\bar{\mathbf{p}}^s$ the PPN generates a parametrization $\bar{\mathbf{t}}^s \subset [0, 1]$. This parametrization is rescaled to $[u_{s-1}, u_s]$ and adapted to the sub-/supersampling of $\bar{\mathbf{p}}^s$, yielding $\mathbf{t}^s$. We insert a parametric value

$$t_i = t_\alpha^s + (t_\omega^s - t_\alpha^s)\frac{\text{chordlen}(p_\alpha^s, p_i)}{\text{chordlen}(p_\alpha^s, p_\omega^s)}, \tag{3.2}$$

to $\bar{\mathbf{t}}^s$ for each point p_i that was removed from $\mathbf{p}^s$ in subsampling. In 3.2 chordlen is the length of the polygon defined by a point sequence and p_α^s and p_β^s are the closest neighbors of p_i to the left and right in the subsampled segment with parameters t_α^s and t_β^s. Parameters t_i^s corresponding to placeholder points are removed from $\bar{\mathbf{t}}^s$. An initial knot vector for the initialization of the parametrization is required. This knot vector is computed by first defining $u_0 = 0$ and $u_n = 1$. Then, for each segment, except the last segment, one knot u_i is added:

$$u_i = u_{i-1} + \frac{\text{chordlen}(\mathbf{p}^s)}{\text{chordlen}(\mathbf{p})}, \quad i = 1, ..., r - 1.$$

This yields a start- and end-knot for every point sequence segment.

Refinement

Compared to the SVM-based method of Section 3.2 this approach applies knot vector refinement and is thus not able to instantly yield a knot vector. In the refinement step, additional knots are added to point segments with the largest approximation error. After an intermediate approximation with the current knot vector, the segment $\mathbf{p}^s$ with largest HD to this approximation is determined. For a given $\bar{\mathbf{p}}^s$ and $\bar{\mathbf{t}}^s$ a new knot $\bar{u}_s \in [0, 1]$ is predicted by the KSN. This knot is then mapped to the actual knot value range $[u_{s-1}, u_s]$ by

$$\tilde{u}_s = u_{s-1} + \bar{u}_s(u_s - u_{s-1}). \tag{3.3}$$

Instead of inserting $\tilde{u}_s$ into the knot vector the parameter t_i closest to $\tilde{u}_s$ becomes part of $\mathbf{u}$. This correction of $\tilde{u}_s$ ensures that (2.27) is satisfied.[4] The refinement proceeds until a desired number of knots is reached, or a specific error threshold is satisfied.

3.3.2 Neural Network Architecture

Figure 3.6 gives an overview of the two interdependent neural networks used in our approach. As we will show in this chapter, we can solve

---

[4]Note that $\tilde{u}_s$ could be inserted to $\mathbf{u}$ directly, as long as (2.27) is satisfied.

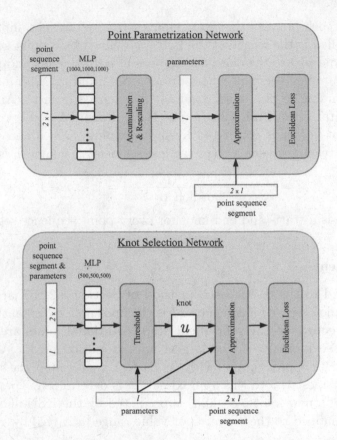

Figure 3.6: The two interdependent neural networks for point parametrization and knot selection. The different entities of the architecture are explained in Section 3.3.2

the absence of a parametrization ground truth by including B-spline approximation as an active part of the neural network architectures. Since the neural networks process, one segment p^s at a time the upper index s and the over-bar are dropped in notation. The neural networks can process any two-dimensional ordered point set as long as the training set complexity does not differ too much from the point set $\mathbf{p}$ for approximation.

Point Parametrization Network

For a point set $\mathbf{p} = (p_0, \ldots, p_{l-1})$ a parameter vector $\mathbf{t} = (t_0, \ldots, t_{l-1})$ consisting of parametric values $t_i = t_{i-1} + \Delta_{i-1}$ is defined. We propose to train a point parametrization network to predict Δ_{i-1}. Given an ordered set of input points $\mathbf{p} = (x_0, ..., x_{l-1}, y_0, ..., y_{l-1})$ with point coordinates (x_i, y_i) for some point p_i the PPN is trained to predict $\Delta = (\Delta_0, ..., \Delta_{l-2})$ with

$$\Delta_i > 0, i = 0, \ldots, l - 2, \tag{3.4}$$

in the parameter domain $u_0 = t_0 = 0$ and $u_n = t_{l-1} = 1$. The parameters need to satisfy $t_0 < t_1$ and $t_{l-2} < t_{l-1}$. $\mathbf{p}$ is given as input to a MLP which maps to the MLP output $\Delta^{mlp} = (\Delta_0^{mlp}, ..., \Delta_{l-2}^{mlp})$ of size $l - 1$.

Accumulation & Rescaling Layer: Based on Δ^{mlp} a parameter vector $\mathbf{t}^{mlp}$ with $t_0^{mlp} = 0$ and

$$t_i^{mlp} = \sum_{j=0}^{i-1} \Delta_j^{mlp}, i = 1, \ldots, l - 1,$$

is computed. Since, t_{l-1}^{mlp} is not necessarily 1, rescaling of $\mathbf{t}^{mlp}$ to $[0, 1]$ resulting in $\mathbf{t}$

$$t_i = t_i^{mlp} / \max(\mathbf{t}^{mlp}),$$

is performed. We ensure positivity of Δ^{mlp} by applying the softplus activation function

$$f(x) = \ln(1 + e^x)$$

for neurons of the MLP in the PPN, so that (3.4) is satisfied.

Approximation Layer: As a consequence of missing ground truth, we include the B-spline approximation directly into the network. This way we can optimize network parameters concerning the approximation quality resulting from a parametrization generated by the MLP. Based on points $\mathbf{p}$ and rescaled parametric values $\mathbf{t}$ as well as an empty knot vector $\mathbf{u} = (0, 0, 0, 0, 1, 1, 1, 1)$ a B-spline curve of degree $k = 3$

is computed. Since $\mathbf{p}$ might be a point segment and thus part of a larger point set we approximate without endpoint interpolation (2.26). The approximation layer's output $\mathbf{p}^{app} = (p_0^{app}, \ldots, p_{l-1}^{app})$ is the approximating B-spline curve evaluated at $\mathbf{t}$. For backpropagation the derivative given in (2.22) is used.

Euclidean Loss: The loss function is defined as

$$\frac{1}{l} \sum_{i=0}^{l-1} \|p_i - p_i^{app}\|_2. \tag{3.5}$$

Knot Selection Network

For a given set of points $\mathbf{p}$ and a parametrization $\mathbf{t}$ computed by the PPN the KSN predicts a new knot value u in the interval $[0,1]$. In the refinement setting the KSN will be applied to segments that need further refinement (see Section 3.3.1). Its input size is $3l$. An MLP is applied to map the input to a single output value u^{mlp}. We use the ReLU activation function [33] for intermediate layers and the Sigmoid activation function [41] in the output layer.

Threshold Layer: The knot predicted by the MLP has to satisfy $u \in (0,1)$, and $\mathbf{t} \cap [0, u] \neq \emptyset$ and $\mathbf{t} \cap [u, 1] \neq \emptyset$, to ensure (2.27). A threshold layer is applied to map u^{mlp} to

$$u = \begin{cases} \varepsilon & , \text{if } u^{mlp} \leq 0 \\ 1 - \varepsilon & , \text{if } u^{mlp} \geq 1 \\ u^{mlp} & , \text{otherwise.} \end{cases}$$

To ensure that knot multiplicity stays equal to k we introduce a small $\varepsilon = 1e - 5$. The resulting u corresponds to $\bar{u}$ in (3.3) where the knot is mapped to its actual knot value range.

Approximation Layer: This layer is equal to the Approximation Layer of the PPN with the exception that the knot vector becomes $\mathbf{u} = (0,0,0,0,u,1,1,1,1)$. For backpropagation, the derivative given in

(2.23) is used. The output of this layer are points $\mathbf{p}^{app} = (p_0^{app}, \ldots, p_{l-1}^{app})$ evaluated at $\mathbf{t}$ for the approximated B-spline curve.

Euclidean Loss: The loss function is identical to that of the PPN.

3.3.3 Training Process

Much like in Section 3.2.2 a diverse training data set is needed. Compared to the training and test data sets for the SKP method the deep learning parametrization approach (PARNET) does not require knot vector computation by exhaustive search. For SKP the exhaustive search limited the number of training samples. Since this limitation does not apply to PARNET, we propose to compute a larger set of curves while using the same method described in Section 3.2.2.

We generate random control points c_i to compute B-spline curves of degree $k = 3$ with $(k+1)$-fold end-knots and no interior knots. For the y-coordinates, we use $\sigma = 2$ and $\mu = 10$. For the x-coordinates, we use $\sigma = 1$ and $\mu = 10$ for the first control point and increase μ by $\Delta\mu = 1$ for all consecutive control points. Again curves with self-intersections are discarded. With this approach, a data set consisting of 150,000 curves is generated. Then l points $\mathbf{p} = (p_0, \ldots, p_{l-1})$ along each curve are sampled. These curves tend to have increasing x-coordinates from left to right. We also add index-flipped versions of the point sequences to the data set resulting in 300,000 point sequences. 20% of this data set is used as validation data in the training process. The training process consists of the following steps:

1. Train the PPN on the training data set.

2. Discard the Approximation Layer and the Euclidean Loss layer. The Accumulation and Rescaling Layer becomes the network output layer.

3. Predict parametric values $\mathbf{t}$ for the training dataset with the PPN.

4. Train the KSN on the training dataset together with **t** from step 3.

5. Remove the Approximation Layer and the Euclidean Loss layer from the KSN. The Threshold Layer becomes the output layer of the KSN.

The MLPs of the PPN and KSN each consist of three hidden layers with size $1,000$ for the PPN and size 500 for the KSN. We apply dropout [43] to MLP layers and train the networks using the Adam optimizer [48].

3.3.4 Results

In this Section we present approximation results of the PARNET approach. First, we will compare the results of point parametrization to other parametrization methods. Then we will compare parametrization and knot placement results to the DPKP and NKTP method. DPKP and NKTP methods are used in conjunction with the centripetal method for parametrization.

For the evaluation we generated four evaluation sets:

- *Evaluation Set 1* contains 500 curves computed as described in Section 3.3.3. We sample 500 equidistributed (in terms of arc length) points on each curve.

- *Evaluation Set 2* contains the curves from *Evaluation Set 1* but sampled at random parameters.

- *Evaluation Set 3* contains 500 curves computed as described in Section 3.3.3 but with random interior knots without multiplicities. We generate 3 to 8 random interior knots which results in a set of very diverse curves. We sample 500 equidistributed points on each curve.

- *Evaluation Set 4* contains the curves from *Evaluation Set 3* but sampled at random parameters.

Table 3.1: Average Hausdorff distances for approximation without interior knots of the PPN compared to common parametrization methods.

	Evaluation set 1	*Evaluation set 2*
PNN	0.0224	0.0992
Uniform	0.2097	0.2095
Chordal	0.2099	0.2001
Centripetal	0.2098	0.2030
PPN	0.0245	0.1088
uniform	0.2040	0.2102
chordal	0.2042	0.1955
centripetal	0.2040	0.2008

We included *Evaluation Set 2* and *Evaluation Set 4* because many parametrization methods use noise filters before parametrization, like e.g. [88]. These filters result in a smooth set of points (or smooth curvature) but also lead to an uneven distribution of points.

Point Parametrization

In Table 3.1 we compare point parametrizations computed by the PPN with the uniform, chordal, and centripetal parametrization. We compare the methods for equidistributed as well as randomly sampled points regarding the average HD over evaluation sets *1* and *2*. Parametrizations computed by the PPN result in approximations with up to eight times smaller HD for *Evaluation Set 1*. For *Evaluation Set 2*, the results are still two times smaller when compared to the other methods. As can be seen in Figure 3.9 approximations parametrized by the PPN can follow the point sets closely. Figure 3.9a shows a close-up on an example where the PPN results in larger HD when compared to the centripetal method.

Knot Selection

We evaluate the effectiveness of our approach by comparing PARNET to the NKTP, the DPKP and the SKP method from Section 3.2. The methods are compared by average HD over *Evaluatoin Set 3* and *Evaluation Set 4*. Figure 3.10 shows the results of knot placement in a range from 5 to 25 knots. On both sets, the PARNET method produces the tightest approximations. Especially with fewer knots in the range from 5 to 14 knots, our method results in smaller HD. With an increasing number of knots, results of DPKP and our method are very close with a small advantage for our method at 25 knots on *Evaluation Set 3* and for DPKP on *Evaluation Set 4*. Consistent with Section 3.2.4 the SKP method performs better than NKTP and leads to a slightly higher HD for larger numbers of knots when compared to DPKP. The SKP leads to a tighter fit at fewer knots when compared to DPKP. Figure 3.8 shows approximation results by NKTP, DPKP and PARNET for two examples from *Evaluation Set 3*. While NKTP produces very smooth results, it fails in complex regions. The DPKP method can approximate regions of high curvature very well but may also lead to wiggles in these regions. PARNET is able to approximate highly curved regions while results are mostly free of wiggles.

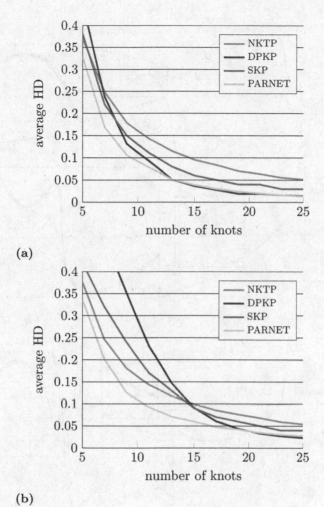

Figure 3.7: Average Hausdorff distance over *Evaluation Set 3* (a) and
Evaluation Set 4 (b) at different numbers of knots for NKTP,
DPKP, SKP, and PARNET

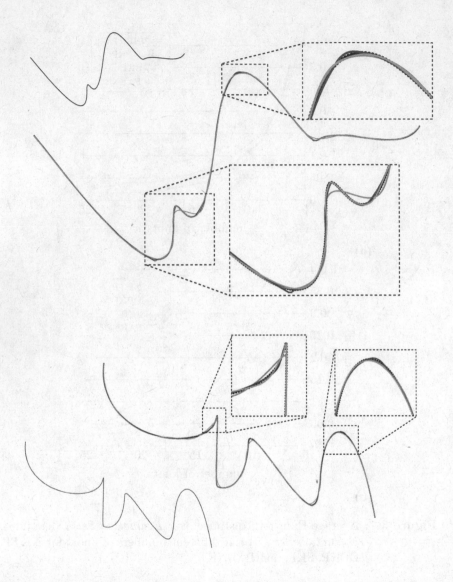

Figure 3.8: Approximations of two point sets by NKTP (orange), DPKP
(blue), and PARNET (green) with a close-up of the dashed
region. Top left and bottom left are the original point
clouds.

3.4 T-spline Surface Approximation

B-spline surface approximation by surface skinning or lofting is the process of approximating a set of contour curves. Each contour curve is computed by approximating a set of sequentially arranged, almost co-planar points. The contour curves require compatible knot vectors. Since individual curves are best approximated with individual knot vectors, making the knot vectors compatible implies either knot insertion or knot removal for specific curve knot vectors. However, for the affected contour curve the additional knots are locally spurious, or the knot removal deteriorates the local approximation. We propose to approximate T-spline surfaces using surface skinning. Approximation by a T-spline surface reduces the number of required control points that would be needed to represent the same surface in B-spline form.

3.4.1 T-spline Surface Skinning

Surface skinning requires a series of $j = 0, \ldots, n$ sequentially arranged point sets p_{ij}, $i = 0, \ldots, m_j$, as shown in Figures 3.12a and 3.11a. Unstructured point sets can be converted to a series of sequential point sets by point cloud slicing, see, e.g. [107, 70].

The points p_{ij}, $i = 0, \ldots, m_j$, are approximated by the j-th contour curve $\mathbf{R}_j(u)$. For the approximation of the contour curves, individual knot vectors for each contour curve yield best approximation results for a minimal number of locally distributed knots. Tensor product B-spline approximation relies on a compatible, global knot vector. Generating a global knot vector from the individual knot vectors requires either knot insertion of locally spurious knots or knot removal increasing the approximation error locally. Following [111], a T-spline surface for skinning is defined as

$$\mathbf{S}(u,v) = \sum_{j=0}^{n} \sum_{i=0}^{n_j} b_{ij} N_{i,k_1}^{(j)}(u) N_{j,k_2}(v) \tag{3.6}$$

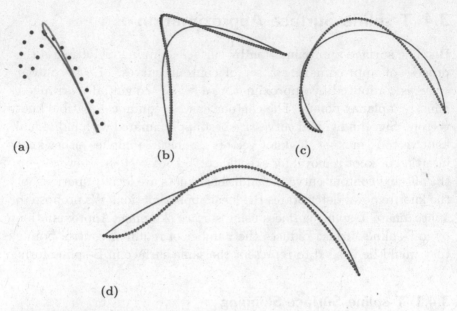

Figure 3.9: Results of parametrizations for examples of *Evaluation Set 1* by the centripetal method (blue) and by the PPN (red), approximated without interior knots.

with control points b_{ij}. The B-splines $N_{i,k_1}^{(j)}$, $i = 0, \ldots, n_j$, of degree k_1 are defined by individual knot vectors

$$\mathbf{u}_j = (u_0^{(j)}, \ldots, u_{n_j}^{(j)}), \quad j = 0, \ldots, n,$$

where the knots $u_0^{(j)} = 0$ and $u_{n_j}^{(j)}$ have multiplicity $k_1 + 1$. The B-splines N_{j,k_2} of degree k_2 are defined by a single knot vector

$$\mathbf{v} = (v_0, \ldots, v_n),$$

where the knots $v_0 = 0$ and v_n have multiplicity $k_2 + 1$ and

$$v_j = v_{j-1} + \|\nabla p_{0,j}\| + \|\nabla p_{m_j,j}\|,$$

for $j = 1, \ldots, n$. If the knot vectors $\mathbf{u}_j$ coincide, the T-spline surface reduces to a tensor-product B-spline surface.

The steps for approximation with a T-spline skinning surface are the same as for a B-spline skinning surface, only without compatible knot vectors. First, the contour curves $\mathbf{R}_j(u)$ are computed as described in Section 2.2.3 by approximation of the point sets $\{p_{ij} : i = 0, \dots, m_j\}$ for $j = 0, \dots, n,$. Second, the T-spline skinning surface is constructed to interpolate these contour curves. Using a set of control curves $\mathbf{Q}_j(u)$ the interpolation constraints are given by

$$\mathbf{S}(u, v_l) = \sum_{j=0}^{n} \mathbf{Q}_j(u) N_{j,k_2}(v_l) = \mathbf{R}_l(u),$$

for $l = 0, \dots, n$, and

$$0 = \left. \frac{\partial^2 \mathbf{S}(u, v)}{\partial v^2} \right|_{v=v_0}, \qquad 0 = \left. \frac{\partial^2 \mathbf{S}(u, v)}{\partial v^2} \right|_{v=v_n}.$$

This yields the $(n + 3) \times (n + 3)$ system of linear equations

$$\mathbf{M} \cdot \begin{pmatrix} \mathbf{Q}_0(u) \\ \mathbf{Q}_1(u) \\ \mathbf{Q}_2(u) \\ \vdots \\ \mathbf{Q}_n(u) \\ \mathbf{Q}_{n+1}(u) \\ \mathbf{Q}_{n+2}(u) \end{pmatrix} = \begin{pmatrix} \mathbf{R}_0(u) \\ \mathbf{R}_0(u) \\ \mathbf{R}_1(u) \\ \vdots \\ \mathbf{R}_{n-1}(u) \\ \mathbf{R}_n(u) \\ \mathbf{R}_n(u) \end{pmatrix}.$$

for the control curves $\mathbf{Q}_j$. Since $\mathbf{M}$ is diagonally dominant, $\mathbf{M}^{-1}$ is well-defined. Thus, the control curves $\mathbf{Q}_j(u)$ are linear combinations of contour curves $\mathbf{R}_l(u)$

$$\mathbf{Q}_j(u) = \sum_{j=0}^{n} q_{ij} \mathbf{R}_j(u),$$

where the coefficient vectors are determined by the rows of $\mathbf{M}^{-1}$. For a detailed description of $\mathbf{M}$ and its inverse $\mathbf{M}^{-1}$ refer to [111].

Finally, $S(u, v)$ is computed from control curves

$$\widetilde{\mathbf{Q}}_j(u) = \sum_{i=0}^{n_j} \widetilde{q}_{ij} N_{i,k_1}^{(j)}(u),$$

which interpolate $\mathbf{Q}_j$ at the knots $\mathbf{u}_j$.
For our experiments we used $k_1 = k_2 = 3$.

3.4.2 SKP and PARNET for T-spline Surface Skinning

After slicing, we assume the contour curves $\mathbf{R}_j(u)$ to have small torsion. We apply principal component analysis and reduce the three-dimensional point data to two dimensions. Using an approximating T-spline skinning surface, the knot vectors $\mathbf{u}_j$ do not need to be compatible. Each $\mathbf{u}_j$ is computed independently using the SKP or PARNET method for the approximation of contour curve $\mathbf{R}_j(u)$.

3.4.3 Results

To evaluate the effectiveness of the SKP and PARNET method for surface skinning, we compare cubic approximations against results computed by the NKTP and DPKP methods. NKTP, DPKP and SKP use centripetal point parametrization. Figure 3.12 shows an ordered point set for surface skinning of a dustpan together with approximation results. This point cloud consists of $7,500$ points. Figure 3.13 shows a close-up of the dustpan handle. The screwdriver handle example in Figure 3.11 consists of $3,000$ points. Figures 3.10b and 3.10a show results for the approximation between 75 and 375 knots for both point sets. For both examples, the SKP method leads to the tightest approximation at 375 knots. Looking at the Figures 3.12, 3.13 and 3.11 the SKP method also leads to a very smooth approximation that follows the point set closely. The PARNET method performs poor on the screwdriver model and is slightly better than the NKTP method on the dustpan model. The visual quality of approximations computed using the PARNET method is unconvincing. It leads to sporadic wiggles of the approximated surface.

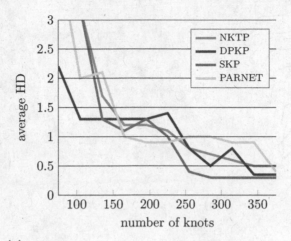

(a)

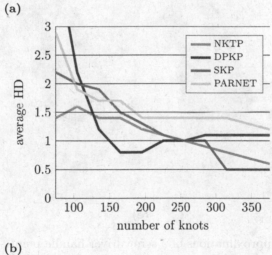

(b)

Figure 3.10: Average HD for the approximation of (a) the dustpan
point set in Figure 3.12a and (b) the screwdriver point
set in Figure 3.11a at different numbers of knots for NKTP,
DPKP, SKP, and PARNET

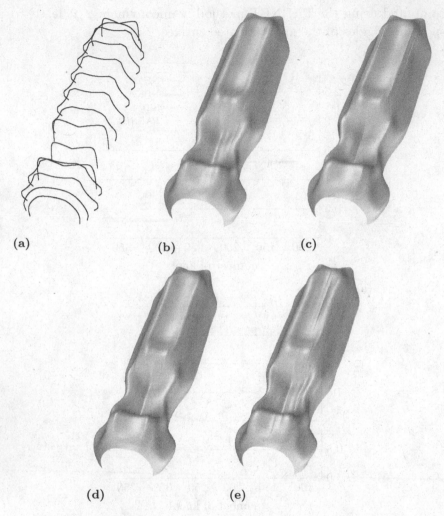

(a)

(b)

(c)

(d)

(e)

Figure 3.11: Approximations of a screwdriver handle point set (a)
computed using parametrizations by NKTP (b), DPKP
(c), SKP (d) and PARNET (e).

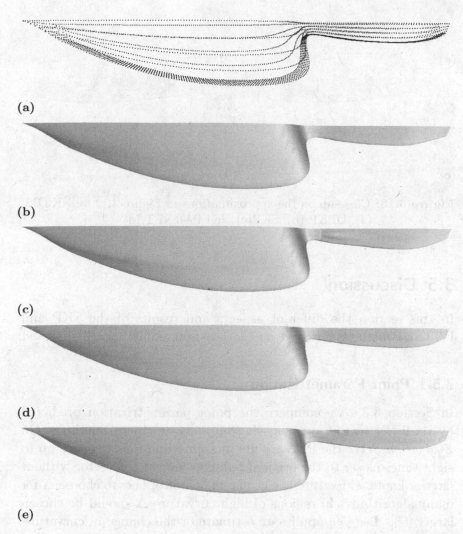

(a)

(b)

(c)

(d)

(e)

Figure 3.12: Approximations of a dustpan point set (a) computed using parametrizations by NKTP (b), DPKP (c), SKP (d), and PARNET (e).

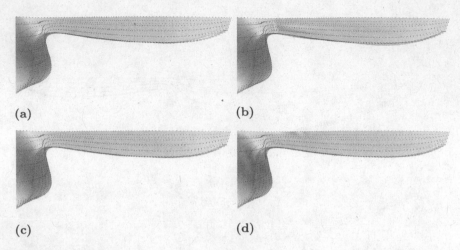

(a)　　　　　　　　　　　　　(b)

(c)　　　　　　　　　　　　　(d)

Figure 3.13: Close-up on the approximations in Figure 3.12 by NKTP (a), DPKP (b), SKP(c), and PARNET (d).

3.5 Discussion

In this section the different aspects and results of the SKP and PARNET methods for curve and surface approximation are discussed.

3.5.1 Point Parametrization

In Section 3.3.4 we compare the point parametrization predicted by the PPN to the uniform, chordal and centripetal method. On *Evaluation Set 1* the PPN results in approximations that are up to eight times closer to the original point set for a knot vector without interior knots. Curvature is a strong indicator of how to choose Δ for many algorithms. In regions of high curvature Δ should be chosen larger [66]. Lee [59] applies an estimate of the change in curvature, the exponent method, to choose Δ. Since the exponent method assumes that regions of high curvature are sampled densely, it fails for equidistributed point sets as can be seen from Table 3.1 and Figure 3.9.

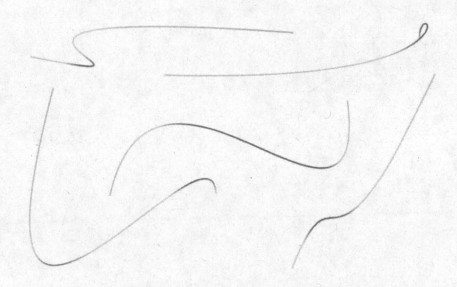

Figure 3.14: Heatmap-colored curves of *Evaluation Set 1* colored by
parametrization value Δ. Blue corresponds to low values
for Δ while red corresponds to large values for Δ.

The parametrizations by the PPN lead to very tight approximations
without interior knots as can be seen in Figure 3.9. In *Evaluation
Set 1*, only one example led to a larger HD for the parametrization
predicted by the PPN compared to the centripetal method. Part of
this example can be seen in Figure 3.9a.

In Figure 3.14 we show different examples of *Evaluation Set 1* in
heatmap coloring for parametric distances Δ. From Figure 3.14 it is
clear that the PPN has learned to predict curvature to some extent.
In regions of high curvature, Δ is large while it is small in straight
segments. There are exceptions, e.g. for inflection points where the
PPN keeps Δ at a higher value. Since points in *Evaluation Set 1* are
equidistributed the PPN cannot base its predictions on point distances
which suggests an implicit understanding of curvature.

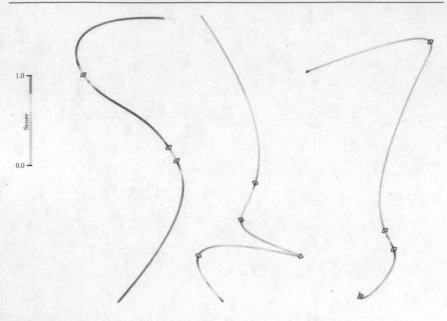

Figure 3.15: Example curves from the test set of Section 3.2 colored by score. High scores are red and low scores are blue. Diamonds represent LCM points.

3.5.2 Knot Selection

Given the score values by the SKP method from Section 3.2 and the knot locations predicted by the KSN it is instructive to analyze the SKP score distribution as well as KSN knot locations. Figure 3.15 shows three curve examples colored by SKP score while Figure 3.16 shows eight curves together with knot locations predicted by the KSN. We will discuss several observations that can be made by looking at these figures.

Refinement

In Figure 3.15 one can see that the score between LCM points changes smoothly. This is a result of a smooth transition of feature

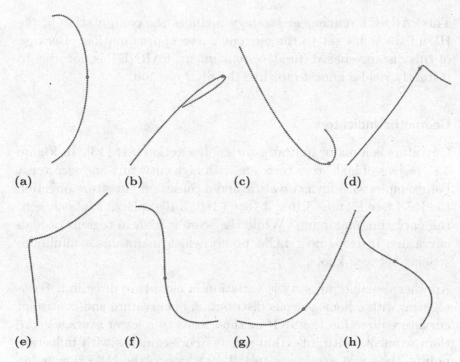

Figure 3.16: Single knot (red) selected by the KSN on point sets of
Evaluation Set 1.

values between LCM points. At LCM points the change in score
is discontinuous which is a result of feature scope. Since many
features are based on point relations inside an LCM segment, segment
transitions lead to score discontinuity. While the score to the left or
right of an LCM point can change significantly, inserting a knot to
either one side may lead to comparable approximation results. A knot
may be shifted within a certain flexibility interval without changing
the approximation result much [75]. Segments with lower average
score result in lower scores at the segment LCMs. This indicates
that the score does not only measure the fitness of a local position
for becoming a knot, but also the fitness of the enclosing segment
to be further refined. This is reflected by our refinement strategy in
Algorithm 1 where scores with maximum score sum are refined first.

The PARNET refinement strategy includes the computation of the HD of the point set to the current curve approximation. Because of this distance-based iterative refinement, PARNET is not able to instantly yield a knot vector like the SKP method.

Geometric Indicators

Curvature is a major indicator for knot selection [81, 113]. In Figure 3.15 regions of high score correlate with high curvature and vice-versa. For examples with high curvature and a consistent curvature direction the KSN (see Figures 3.16b, 3.16c, 3.16d, 3.16e) places knots close to the curvature maximum. While the score is high in regions of high curvature, it drops near LCM points which maintains a minimum spacing between knots.

Another valuable property is variation in curvature direction. For a segment with a homogeneous distribution of curvature and consistent curvature direction the SKP method leads to a lower average score than segments with inflection points. For segments with inflection points, the score maximum usually is close to the inflection point. The same can be observed for knots predicted by the KSN. Inflection points are often favoured as locations for new knots even if there are apparent curvature maximum points in the point set like in Figures 3.16f and 3.16h.

As stated by Razdan in [81] approximation quality benefits from splitting larger segments even if there are no critical features in that segment. This coincides with low average scores of the SKP method in small segments. If there are no significant geometric features, the KSN tends to insert knots close to the point set median. This coincides with the strategy by Park et al. to split segments so that they are of equal complexity [71].

3.5.3 Surface Approximation

A significant benefit of the skinning approach is that knot placement methods for curve approximation can be applied directly to surface approximation. Since the point sets for approximation consist of co-planar contour curves torsion can be disregarded. Surface skinning requires that the contour curves do not deviate too much since this would lead to distortion in-between contour curves [111]. The same holds for the knot vectors of the neighbouring contour curves. Methods that lead to knot vectors which have more substantial deviation among adjacent contour curves lead to lower quality approximations. The average Euclidean distance of adjacent knot vectors of PARNET is up to 7 times larger than for NKTP and 1.5 times larger when compared to SKP or DPKP. While PARNET can outperform NKTP, DPKP and SKP for two-dimensional curve approximations, its performance for surface skinning was unsatisfactory. This is a result of larger variance between knot vectors of adjacent contour curves. It also explains the proper performance of the NKTP method for surface skinning while it performs worst for two-dimensional curves. Point clouds of very complex shapes should be sliced at smaller intervals to compensate for this behaviour.

3.6 Conclusion

In this chapter two machine learning approaches for computing parametrizations for Curve and Surface Approximation are presented. In Section 3.2 a method using SVMs to estimate a score of pre-computed parametric values of points as knot values is proposed. Local score maxima subdivide the initial knot interval. Intervals are further subdivided based on their sum of scores. The trained support vector machines are able to learn characteristics of positions along a curve where knot placement has a positive impact on approximation quality.

In Section 3.3 the PARNET method is introduced. This approach computes both, point parametrizations, as well as knot vectors using interdependent deep neural networks. Since approximation to a specific error threshold requires adding knots in succession, we decoupled point parametrization and knot placement in separate networks. When the number of knots is known upfront PPN, and KSN could be merged into a single network. Our experiments show that neural networks are able to predict parametric values **t** and knots **u** simultaneously. The method results in tight approximations. It works well for unevenly spaced points although we trained on evenly spaced point sequences. The trained neural networks generalize well to previously unseen data with versatile characteristics. The neural networks of the PARNET method directly integrate B-spline curve approximation as part of the neural network architecture. We hope that this will enable others to apply neural networks for approximation-related problems.

For knot placement the two approaches have been compared against each other and the NKTP method for instant knot vector computation, as well as the DPKP method which is an adaptive refinement based method. While PARNET outperforms all other methods when used for B-spline curve approximation, the SKP method leads to tighter approximations in T-spline surface approximation. Due to the support vector machine's ability to generalize well, even on smaller data sets [28], the SKP method outperforms state of the art knot placement methods, using only a moderately sized training set. For neural networks there exists an almost linear correlation between training set size and performance [94]. Then again the PARNET method neither requires exhaustive search of a ground truth knot vector nor the computation of geometric features.

A drawback of the PARNET method is the need for segmentation as well as sub- and supersampling of point sequences together with MLPs of fixed input size. We deliberately chose to segment and sample straightforwardly to show that the approximation quality is not a result of preprocessing but of the parametrization predicted by the neural networks. To be able to process point sequences of an arbitrary size we plan to investigate the application of recurrent neural networks instead of pure MLPs.

One aspect that limits both approaches are the synthetic training data sets. Real world data would be preferable. For neural networks, it is common to retrain the networks for the purpose of specialisation. When applying the PARNET method, we suggest to use the synthetic training data in pre-training. The neural networks may then be subsequently improved with additional real-world data. One of our future efforts will be the generation of large enough real-world data sets.

4 Classification of Geometric Primitives in Point Clouds

As introduced in Section 1.1 classification of geometric primitives in point clouds is an integral part of the reverse engineering pipeline. It is also a vital part in the particular case of separating surface structure from base geometry. Compared to object recognition or object retrieval tasks the classification of geometric primitives is somewhat harder. On the one hand, the data for classification only contains part of the object in question. Most of the time only patches of geometric primitives are used in CAD. On the other hand, the difference between different classes, e.g. spheres and ellipsoids, may be quite small which leads to frequent missclassifications. Fitting a geometric primitive to a point cloud is usually done by one of two approaches:

- Given a list of geometric primitive classes, fit each class to the point cloud and select the primitive that yields the smallest approximation error. Algorithms following this methodology usually apply random sample consensus or RANSAC [89, 69].

- Decide on the class of the primitive first and fit the primitive afterwards.

RANSAC-based methods have two major drawbacks: Testing each primitive class is expensive and noisy point clouds may lead to fitting the wrong primitive class. Consequently, we propose to classify the primitive first explicitly. Since CAD models usually contain only parts of geometric primitives we propose to classify patches of primitives using machine learning.

© Springer Fachmedien Wiesbaden GmbH, part of Springer Nature 2020
P. Laube, *Machine Learning Methods for Reverse Engineering of Defective Structured Surfaces*, Schriftenreihe der Institute für Systemdynamik (ISD) und optische Systeme (IOS), https://doi.org/10.1007/978-3-658-29017-7_4

First, we evaluate the performance of support vector machines for primitive classification within different feature spaces. There exists a multitude of geometric features for the characterization of geometric objects. In literature, only a small subset of these geometric features is used. Therefore an in-depth performance analysis of the known features and feature combinations is given.

Secondly, we compare the performance of the support vector machine classifier to deep neural networks on voxelized point clouds. Figure 4.1 shows the six primitive classes we consider for classification.

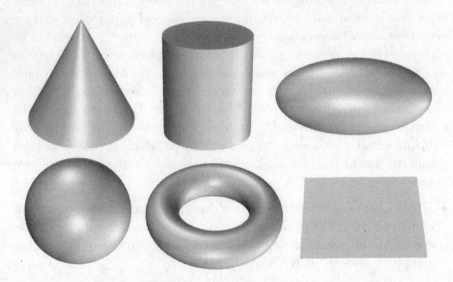

Figure 4.1: The six primitive classes: cone, cylinder, ellipsoid, sphere, torus, and plane.

4.1 Related works

Three-dimensional object recognition is an active field of research and has become popular as a result of the ever increasing number of capturing devices. Technological advancements like autonomous driving require the classification of objects such as trees or humans in scenes [13, 14, 67]. The basic concept of many approaches is the use of geometric features extracted from point relations. Osada et al. introduced the well-known shape distributions [68] in search for a suitable shape signature. Distributions are defined by histograms of angles, areas, distances, and volumes of random point tuples. Another essential feature descriptor is the surflet-pairs feature introduced by Wahl et al. [103]. Surflet-pairs are built upon the relation of pairs of points and their normals. Curvature, as well as curvature direction, has been identified as a quintessential feature in the description of shapes by Hetzel et al. [42]. Rusu et al. [84] propose persistent feature histograms to characterize geometry. The advantage of these feature histograms is that they are independent of point cloud pose and point cloud density.

While thresholding is possible to distinguish classes of objects or primitives from one another [21] modern machine learning methods usually outperform fixed thresholds. Endoh et al. [25] apply locally linear embedding paired with clustering to reduce the number of required training samples for retrieval of three-dimensional models. In [109] Yamamoto et al. examine the performance of supervised, semi-supervised, and unsupervised dimension reduction for object retrieval. Their feature space is limited to the surflet-pairs feature of Wahl et al.

Using support vector machines, Arbeiter et al. [1] classify a small set of geometric primitives as well as corners and edges. They apply features derived from point relations as well as curvature measures. Very recently deep neural networks have been applied to the problem object classification. Especially voxel-based representations have been the focus of research on neural networks for classification. Wu et al. [108] apply the so-called ShapeNets for classification of voxelized models in a large CAD database. The resulting three-dimensional filter kernels strongly resemble their two-dimensional counterparts. Maturana et al. [67] apply a similar concept for real-time object recognition. They propose VoxNet a network architecture integrating a volumetric occupancy grid. In contrast to voxel-based representations Qi et al. [80] apply neural networks to classify and segment shapes in point clouds. While the network performs well on their data set they suffer from the problem of fixed input resolution.

4.2 Training Data Generation

Scans of scenes or objects acquired using laser line scanners comprise a unique structure. While being unorganized regarding point neighbourhood, these point clouds hold a certain systematic structure that is a result of the scanning process:

- Since the point cloud is the result of the laser line projected onto the object; these laser lines are visible in the resulting scan. While points within a laser line have fixed distances the distance of distinct laser lines is arbitrary.

- The scanning strategy will be visible in the scanning result. If the Scanner is human-operated, each scan will have a distinct pattern [17]. Even the result of a machine-operated scanner will strongly depend on the surface in question and the scanning strategy.

Since there are no data sets available to date we propose to create synthetic scans using a laser scanning simulation.

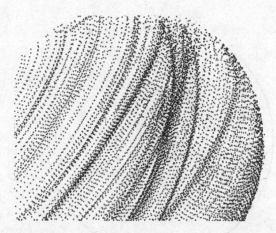

Figure 4.2: Close-up on a point cloud of a sphere scanned with a human-operated laser scanner.

4.2.1 Synthetic Primitive Patches

The simulation is based on ray tracing of a simulated laser line scanner onto geometric primitives. By sweeping fans of rays from a scan position over the geometric primitive, the scanning process is simulated. The position and pose of the laser line probe is defined by two spheres around the centroid of the geometric primitive. Scan positions and poses are defined by a random points p_1 on the outer sphere, a view direction $s_1 = p_2 - p_1$ corresponding to a random point p_2 on the inner sphere, and a random view-up vector $s_2 \perp s_1$. The plane perpendicular to s_2 contains the reference fan, which is a cone of rays with aperture angle φ_1 and axis s_1. The reference fan contains φ_1/φ_3 equidistant rays. To generate additional fans at p_1 the reference fan is rotated around $s_3 = s_2 \times s_1$ by φ_2-angle steps. p_1, p_2 and s_2 are uniformly distributed. This setup is shown in Figure 4.3.

Two types of systematic errors are included: An error offset is applied to the scanner position p_1 as well as each scanned point. Both errors are normally distributed with zero mean and standard deviation of

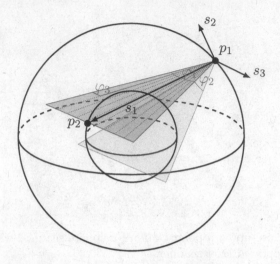

Figure 4.3: The setup of the simulation of a laser scanner.

0.0125 and 0.0025, respectively.

The training data is generated by extracting point cloud patches $\mathbf{p}_j$ from a simulated scan of a complete primitive. Each point cloud patch is defined to contain all points of the point cloud within a predefined cube, of edge length 0.7, centred at a random point of the point cloud. Patches with less than 150 points are discarded. See Figure 4.4 for examples of point cloud patches. This way 9,600 point cloud patches are created distributed equally among the six primitive classes. 80% of the data set is used for training the rest is used for evaluation by true-positive-rate. Compared to training sets for deep learning models this number is small. We limit our data set to 9,600 samples for two reasons:

- The training complexity of support vector machines is at least $\mathcal{O}(n^2)$ [6] where n is the number of training samples. So the number of training samples should stay rather small.

- Due to the nature of the support vector machine model, a larger training set will not necessarily yield an improved classifier.

The geometric parameters for the primitives are normally distributed with mean and standard deviation as given in Table 4.1.

Table 4.1: Properties of the normal distribution of primitive parameters.

Primitive parameter	Mean	Standard deviation	Primitive parameter	Mean	Standard deviation
Plane radius	3.0	0.5	Sphere radius	4.0	2.0
Cylinder height	6.0	0.5	Ellipsoid radius x	4.0	0.4
Ellipsoid radius y	2.0	0.4	Ellipsoid radius z	2.0	0.4
Cylinder radius	1.0	0.5	Torus radius	3.0	0.3
Cone height	6.0	0.5	Torus tube radius	0.8	0.2
Cone radius	1.0	0.5			

Additionally the following affine transformations are applied to each primitve:

- Random rotation about each axis in the range $[0, 2\pi]$

- Random scaling in the range between $[0.9, 1.1]$

To become independent of the point cloud density we homogenize point cloud patches. The density d_P of a patch $P = \{p_1, \ldots, p_n\}$ is computed as

$$d_P = \frac{1}{n} \sum_{i=1}^{n} \sum_{j=1}^{k} \frac{\delta(p_i, q_j)}{k},$$

where q_j are the k-nearest neighbors of p_i and $\delta(p_i, q_j)$ is the Euclidean distance. Based on d_P and the target density d_t, a scaling factor $s = d_t/d_P$ is computed. Each point of P is scaled with s. We used $d_t = 0.01$. Of course, this homogenization needs to be applied to every point cloud before classification by the trained support vector machine.

4.2.2 Volumetric Grid

Modern convolutional neural networks require either a fixed input size or an input space in which localized convolutions can be defined. To

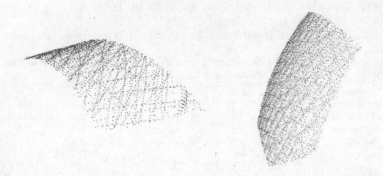

Figure 4.4: Two patches of the synthetic data set for primitive
        classification of a cylinder (left) and a sphere (right).

be able to process an arbitrary number of points using neural networks
we convert point cloud patches to a volumetric grid of voxels $\mathbf{V}$. Before
voxelization, the point cloud patches are translated to the origin and
scaled so that they are fully contained inside the range $[-1, 1]$ for all
coordinates. Voxel grids are then computed by subdividing the space
in the range $[-1, 1]$ for all coordinates by voxel grid dimensionality
d. In our case $d = 64$. Each voxel is assigned the number of points it
contains. The voxel grid is then normalized. Each voxel $\mathbf{V}(i, j, k)$ is
normalized to

$$\mathbf{V}(i, j, k) = \frac{\mathbf{V}(i, j, k) - \mathbf{V}_{min}}{\mathbf{V}_{max} - \mathbf{V}_{min}}$$

where $\mathbf{V}_{min}$ is the minimum of voxel grid $\mathbf{V}$ and $\mathbf{V}_{max}$ its maximum.
Figure 4.5 shows a sketch of a voxel grid.

Since the performance of neural networks is strongly related to the
number of training samples a data set of 36,000 patches for training
is computed as described in Section 4.2.1. These patches are then
voxelized to yield the final volumetric data set.

4.3 Geometric Features for SVM-based Classification

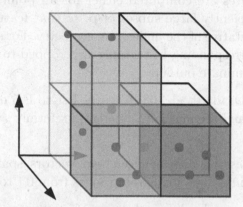

Figure 4.5: Sketch of a 2x2 voxel grid. Color-intensity corresponds to the number of points containted in each voxel.

For most applications, the computation of features derived from raw data improves the performance of the learned model. As for feature extraction, the goal is to derive values that best explain the underlying data without being redundant. Feature representations not only can reduce the dimensionality of the input space but also incorporate expert domain knowledge from the feature selection process. Another problem is that most machine learning methods, including support vector machines, require a fixed input size while the number of points in a point cloud may vary. In the case of geometric primitives the class information for point cloud data should be derivable from features extracted from the local geometry of the point cloud.

In this Section, we present the geometric features of which some have been used in the past for object retrieval and object classification. These values are arranged in feature histograms with varying numbers of bins for each feature.

Support vector machines have shown to perform well in high dimensional space when used with histogram features [11]. We normalize histogram bins to the value range I. The concrete values for the normalization interval I and the number of bins b are shown in Table 4.2. The geometric features are computed either for all points p_i of a point cloud or a sufficiently large subset of $\mathbf{p}$, so that feature histograms become representative of the underlying geometry. For some geometric features, the corresponding histograms are cropped to the $[0.05, 0.95]$ percentile to eliminate outliers.

Introduced by Osada et al. [68] features F.1 to F.5 depend only on point locations and are mutually different, uniformly sampled random points from $\mathbf{p}_j$:

F.1 *Point angles:* Angles between two vectors spanned by three random points. The histogram ranges from $0°$ to $180°$.

F.2 *Point distances:* Euclidean distance δ between two random points.

F.3 *Centroid distances:* Euclidean distance of random points to the bounding box centroid. The centroid is defined as

$$c = \frac{1}{n} \sum_{i=1}^{n} p_i,$$

where $\mathbf{p}_j$ is a point cloud patch containing n points.

F.4 *Triangle areas:* Square root of the triangle areas of random three-tuples of points. Triangles are defined as

$$A = \frac{1}{2} ab \sin \gamma,$$

with sides a and b enclosing angle γ.

F.5 *Tetrahedron volumes:* Cubic root of the tetrahedron volume V of four random points $p_1, \ldots, p_4$

$$V = |(p_1 - p_4) \cdot ((p_2 - p_4) \times (p_3 - p_4))|/6.$$

F.6 *Cube cell count:* Number of points contained in 8 equally sized cells, which result from a uniform subdivision of the point cloud's bounding box. This feature is not invariant to rotation.

F.7 *K-Median points:* This is strongly related to the well known k-means algorithm. The k-medians algorithm minimizes the sum of squared distances of arbitrary points to randomly initialized k median points until it converges to the final median points resulting in k clusters. From the x, y, and z coordinates three coordinate-histograms are computed and concatenated. $k = 32$ clusters are used.

Following, we list further features that do not solely depend on point locations.

F.8 *Normal angles:* Angles between two normals at two random points.

F.9 *Normal directions:* Coordinates of unit normals at all points. Like for feature F.7 the corresponding feature histogram is the concatenation of three coordinate-histograms.

To estimate the normal at point p in the point cloud, the set B of p's k-nearest neighbors is determined. Here, we used $k = 100$. The principal component analysis of B yields the covariance matrix, whose eigenvector n_p corresponding to the smallest eigenvalue is used to estimate the normal at point p.

Furthermore a set of features depending on curvature are used:

F.10 *Principal curvatures* κ_1, κ_2 are computed by polynomial fitting of osculating jets as in [10].

F.11 *Mean curvatures:* $H = \frac{1}{4}(\kappa_1 + \kappa_2)$.

F.12 *Gaussian curvatures:* $K = \kappa_1 \kappa_2$.

F.13 *Curvature ratios:* $|\kappa_1/\kappa_2|$.

F.14 *Curvature changes:* Absolute difference between a random point's principal curvatures and those of its nearest neighbour. The result is two concatenated histograms.

F.15 *Curvature angles:* Angles between the two corresponding principal curvature directions $\mathbf{v}_1$, $\mathbf{v}_2$ at two random points.

F.16 *Curvature directions:* Coordinates of the two normalized principal curvature directions $\mathbf{v}_1$, $\mathbf{v}_2$ at all points. Thus, the corresponding feature histogram is the concatenation of six coordinate-histograms.

F.17 *Curvature differences:* Absolute differences of the principal curvatures, the Gaussian curvature, and the mean curvature at two random points, optionally weighted by distance.

F.18 *Shape index* as defined in [50] for $\kappa_1 > \kappa_2$

$$S_I = \frac{1}{2} - \frac{1}{\pi} \arctan \frac{\kappa_2 + \kappa_1}{\kappa_2 - \kappa_1}.$$

To combine the classification capabilities of individual geometric features, they can be combined into more general features. In [103] a combined normal based feature of two surflet pairs is proposed. These surflet pairs are defined as point-normal-pairs $(p_1, \mathbf{n}_1)$ and $(p_2, \mathbf{n}_2)$ with normalized normals $\mathbf{n}_1, \mathbf{n}_2$. From two surflet pairs a local, right-handed, orthonormal frame is computed

$$\mathbf{u} = \mathbf{n}_1, \quad \mathbf{v} = ((p_2 - p_1) \times \mathbf{u}) / \left| (p_2 - p_1) \times \mathbf{u} \right\|,$$
$$\mathbf{w} = \mathbf{u} \times \mathbf{v}.$$

This frame yields three geometric attributes

$$\alpha = \arctan(\mathbf{w} \cdot \mathbf{n}_2, \mathbf{u} \cdot \mathbf{n}_2), \quad \beta = \mathbf{v} \cdot \mathbf{n}_2,$$
$$\gamma = \mathbf{u} \cdot (p_2 - p_1) / \|p_2 - p_1\|.$$

Together with the point distance δ, these attributes define the surflet pair feature:

F.19 *Surflet pairs:* The tuple $(\alpha, \beta, \gamma, \delta)$ for two random points.

Further combined features can be constructed by concatenation of their respective histograms. Although any combination of the above features is possible, we only combine those features that prove most effective as individual features.

F.20 *Triple combination* of the best point-, normal-, and curvature-features: F.5, F.8, and F.15.

F.21 *Simple surflet combination* of F.5, F.8, and F.19.

F.22 *Extended surflet combination* of F.19 and F.20.

F.23 *All features combination* of features F.1,...,F.19.

To compute the geometric features often random point pairs, triplets, or quadruplets need to be chosen. In these cases, 2^{17} feature values are sufficient to yield stable feature histograms.

4.4 Neural Network Architecture

In addition to classification using support vector machines and the set of geometric features introduced in Section 4.3 we train a convolutional neural network for classification. As introduced in Section 4.2.2 we will use a volumetric grid of voxels as network input. While this representation is missing many of the details of the original point cloud, it provides a fixed input resolution as well as localized convolutions. To apply convolutions to the volumetric grid, we need to define three-dimensional convolutions. Very similar to (2.16), convolutions of a volumetric grid V with a filter kernel F in three-dimensional space are defined by

$$(\mathbf{V} * \mathbf{F})(i,j,k) = \sum_l \sum_m \sum_n \mathbf{V}(l,m,n)\mathbf{F}(i-l,j-m,k-n).$$

Like for convolution the concept of pooling needs to be adapted. This adaptation is made by simply applying the pooling operation in a three-dimensional neighbourhood.

The network architecture for primitive detection is shown in Figure 4.6 [1]. The architecture consists of three convolutional layers *conv1*,

---

[1] The number of boxes of the convolutional kernels in Figure 4.6 does not represent the true number of kernels used in the network architecture.

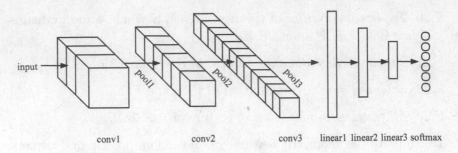

Figure 4.6: Network architecture for primitive classification.

conv2, and *conv3* with kernel sizes 9, 5 and 3 as well as a filter kernel count of 8, 16, and 32. Max pooling is applied after each convolutional layer. Convolutional layers are followed by three linear layers *linear1*, *linear2*, and *linear3* with 1024, 512, and 6 neurons. The ReLU activation function [33] is used at each layer. As network loss, we use the cross-correlation loss together with softmax activation as described in Section 2.1.2.

4.5 Results and Discussion

In this section we outline model selection followed by results for primitive classification using geometric features with support vector machines. An evaluation of support vector machine classification on patches of real scans in comparison to neural networks is presented in Section 4.5.2.

For model selection, a grid search is performed. The same (C, γ)-set and kernel function as described in Section 3.2.4 are applied. Given the 6-class training data the one-versus-all approach trains 6 binary SVMs.

4.5.1 Evaluation of Geometric Features

In the following, the results of individual and combined features for primitive recognition are presented regarding true-positive rates (TPR). Trained support vector machines are evaluated using the test set introduced in Section 4.2. Table 4.2 shows the true-positive-rates for all features and corresponding values for C, γ, b, I, and a flag p, if the histogram is cropped to the $[0.05, 0.95]$ percentile. Note, that for F.23 the histograms are a combination of the respective feature histograms, which are normalized and cropped individually.

The results for the geometric features of Section 4.3 categorized as point-based, normal-based, curvature-based, and combined features are discussed separately. For observations of feature performance when classifying individual primitives feature confusion matrices in Table 4.7 are consulted.

The true-positive-rate of 0.588 of the tetrahedron volumes feature scores highest among the point-based features. While performing well for planes, it confuses ellipsoids with spheres and tori with cylinders and has a weak performance for cone classification. The reasons for this seems to be that point-based features do not capture curvature information sufficiently.

The normal angles feature F.8 has a classification rate of 0.529 and the highest classification rate among normal-based features. It is weak for classifying cones and cylinders. F.8 favours tori so that there is a considerable misclassification of cylinders as tori.

With 0.513 the curvature angle feature F.15 has the highest classification rate of the curvature-based features. It best classifies planes and has the weakest performance for ellipsoids, spheres and cylinders. Confusion of these primitive types can be observed for all features since they have strong geometrical relations, especially when examined locally with noise.

Table 4.2: Test results as true-positive rate (TPR) for all features of the simulated data set.

Feature Histogram	TPR	C	γ	b	I	p
F.1 Point angles	0.564	50	2	64	$[0,1]$	n
F.2 Point distances	0.437	10	5	64	$[0,1]$	n
F.3 Centroid distances	0.370	100	0.1	64	$[0,1]$	n
F.4 Triangle areas	0.515	10	10	64	$[0,1]$	n
F.6 Cube cell count	0.420	5000	0.1	64	$[0,1]$	n
F.7 K-Median points	0.240	1	0.1	96	$[0,1]$	n
F.5 Tetrahedron vol.	0.588	1000	0.75	64	$[0,1]$	n
F.8 Normal angles	0.529	10	1.5	64	$[0,1]$	n
F.9 Normal directions	0.427	1	0.25	96	$[-1,1]$	n
F.10 Principal curv.	0.308	0.1	1.5	128	$[0,1]$	y
F.11 Mean curv.	0.280	5000	10	64	$[0,1]$	n
F.12 Gaussian curv.	0.250	0.01	1	64	$[0,1]$	n
F.13 Curv. ratio	0.286	0.1	2	64	$[0,1]$	n
F.14 Curv. change	0.295	0.01	0.75	128	$[0,1]$	n
F.15 Curv. angles	0.513	15	0.25	128	$[0,1]$	n
F.16 Curv. directions	0.429	5	0.1	192	$[-1,1]$	n
F.17 Curv. differences	0.351	0.1	5	128	$[0,1]$	y
F.18 Shape index	0.280	0.1	2	64	$[0,1]$	n
F.19 Surflet pairs	0.695	5000	0.1	128	$[0,1]$	n
F.20 Triple combi.	0.632	1000	0.01	256	$[0,1]$	n
F.21 Sim. surflet combi.	**0.725**	1000	0.1	256	$[0,1]$	n
F.22 Ext. surflet combi.	0.711	5000	0.01	384	$[0,1]$	n
F.23 All features combi.	0.369	10	0.01	1728	n/a	n/a

The surflet pair feature F.19 performs best for planes, cylinders, and spheres, and has better performance for cones, ellipsoids, and tori than F.5, F.8, or F.15.

The simple surflet combination feature F.21 performs best among all described features. It discriminates spheres, ellipsoids, cylinders, and tori best.

	F.5 Tetrahedron volumes						F.8 Normal angles					
	Cone	Plain	Cyl.	Ellips.	Sphere	Tori	Cone	Plain	Cyl.	Ellips.	Sphere	Tori
Cone	0.19	0.4	0.09	0.15	0.3	0.07	0.3	0.36	0.08	0.17	0.21	0.06
Plain	0.02	1	0	0	0.07	0	0.03	1	0	0.05	0.01	0
Cyl.	0.02	0	0.47	0.18	0.01	0.47	0.09	0.01	0.35	0.2	0.06	0.42
Ellips.	0.02	0.06	0.09	0.61	0.26	0.11	0.08	0.1	0.09	0.53	0.17	0.17
Sphere	0.08	0.17	0	0.12	0.71	0	0.11	0.18	0.03	0.28	0.39	0.07
Tori	0.02	0	0.22	0.07	0	0.84	0.03	0	0.11	0.01	0.08	0.91

	F.15 Curvature angles						F.21 Simple surflet comb.						F.19 Surflet pairs					
	Cone	Plain	Cyl.	Ellips.	Sphere	Tori	Cone	Plain	Cyl.	Ellips.	Sphere	Tori	Cone	Plain	Cyl.	Ellips.	Sphere	Tori
Cone	0.32	0.31	0.1	0.16	0.17	0.11	0.58	0.22	0.1	0.06	0.13	0.02	0.67	0.15	0.11	0.06	0.09	0.02
Plain	0.05	1	0	0.02	0.01	0	0.03	1	0	0	0	0	0.02	1	0	0	0	0
Cyl.	0.40	0.02	0.41	0.18	0.03	0.45	0.05	0	0.74	0.13	0.01	0.16	0.04	0	0.77	0.09	0	0.17
Ellips.	0.09	0.14	0.16	0.41	0.14	0.2	0.05	0	0.06	0.7	0.16	0.11	0.03	0	0.06	0.74	0.15	0.09
Sphere	0.14	0.13	0.08	0.22	0.4	0.08	0.13	0.34	0	0.2	0.65	0	0.12	0.01	0	0.13	0.74	0
Tori	0.04	0.01	0.17	0.06	0.04	0.82	0.01	0	0.2	0.06	0	0.82	0.01	0	0.2	0.07	0	0.8

Figure 4.7: Normalized confusion matrices for features F.5, F.8, F.15, F.19, and F.21 in heat-map coloring.

Comparing results we presented in [9] it is clear that homogenizing the density of the synthesized data decreases classification performance on the synthetic evaluation set by about $10\% - 20\%$. However, we will show that homogenization improves classification results when applied to real scans in Section 4.5.3.

4.5.2 Comparison to Convolutional Neural Networks

The neural network for primitive classification is trained on the training data set defined in Section 4.2.2. The evaluation set introduced in

Section 4.2.1 is voxelized and used for validation. For training, the Adam-optimizer [48] with a learning rate of 1e−3 is applied. Dropout is applied to the linear layers with a dropout rate of 50%. The network is trained using early stopping with a batch-size of 8. On the evaluation set, the neural network achieves a TPR of 0.989. The TPR of the neural network is about 17% higher than that of the support vector machine trained on feature F.21 which performs best among all features. The conversion to volumetric grid is a reduction of input resolution. Even on this lower resolution input, the neural network manages to better distinguish between primitive classes of the test set.

To compare the performance of both methods on real data, three scenes containing different geometric primitives have been scanned. Figure 4.8 shows the different test scenes. Scans were acquired using a FARO Edge ScanArm with laser line probe. To extract patches from point clouds of Figure 4.8 a smoothness based region growing algorithm for segmentation is applied. The implementation in [85] has been used. The total number of resulting point cloud patches is 313. This test set has been labelled with the corresponding primitive class manually.

For classification with neural networks, these patches are voxelized. Geometric features of Section 4.3 are extracted for classification by support vector machines. Since the class distribution in the evaluation set is unbalanced we first determine the TPR for each class separately, sum the results, and divide by the number of classes.

For the test set the neural network arrives at a TPR of 0.177 and the SVM trained on F.21 arrives at a TPR of 0.289. While the neural network outperforms the SVM approach on the synthetic evaluation set, it performs worse on the test set. This performance gap is an indicator that the neural network is not able to generalize well using the training data set. Early stopping and dropout had no major impact on this result.

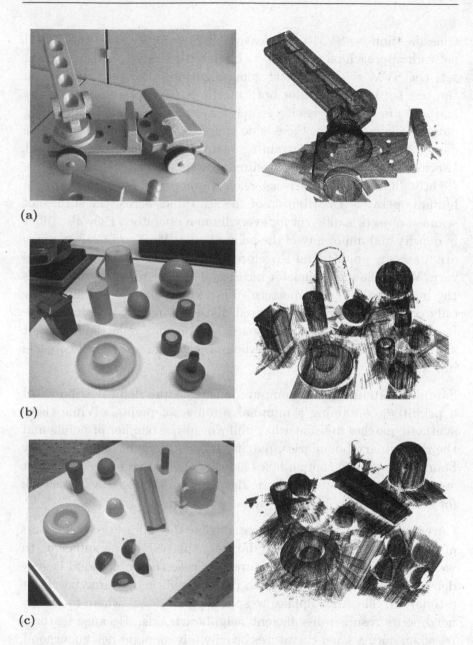

Figure 4.8: Three test scenes (a)-(c) of different objects on the left, and their respective 3d-scans on the right.

Classification by SVM using feature F.21 is superior to the neural network approach on the test set. Even with a much smaller training set, the SVM achieves better generalization. The performance on the test set is still poor for both methods. The presented machine learners are able to classify geometric primitives, but the training data seems inept. Since there is no simple way to collect and produce enough original scans for training - simulated scans had to be used. However, the presented scan simulator generates scans that have slightly different characteristics when compared to scans created by a human operator. Distribution of the scan lines across the surface of scanned objects is different for every human operator. They also differ in density and uniformity of the point clouds. Most real laser scanners are manually guided, and therefore the distances between scan lines vary. While the scan simulator incorporates some randomness, it lacks the human-specific scan-pattern. The error model of our simulator only covers errors in the measured distance and the position of the laser or the camera. Errors caused by specular reflections are not simulated but can occur for polished objects like some of the objects in Figure 4.8.

Sampled patches where computed using a pre-defined window of a primitive containing a minimal number of points. While these synthetic patches are somewhat uniform in the number of points and their imaginary boundaries, patches of real scans have an arbitrary boundary and may contain holes like the one in Figure 4.9. As a first step in creating a more realistic data set we proposed a framework for scanning CAD objects in virtual reality in [17].

Curvatures and normals are estimated using the 100-nearest neighbours of a point. For the synthetic dataset, this value was sufficient to yield stable normals and curvatures. In case the point cloud is very dense, more points are needed to cover a sufficiently large patch for estimation. The same applies to sparse point clouds where too many neighbours result in insufficient neighbourhoods. Because features based on normals and curvatures heavily rely on point neighbourhood, they are not scale invariant.

Figure 4.9: Example point cloud patch of a plane after segmentation of the scene in Figure 4.8b.

In our opinion, these factors contribute to the poor performance of the machine learners on real 3d-scans.

4.5.3 Homogenization

Since point cloud density is a result of the scanning process, the type of 3d-scanner has a significant influence on feature values. Adapting the density of all samples to be uniform before feature computation leads to a better generalization of the approach. To evaluate the impact of homogenization on classification results, support vector machines have been trained on non-homogenized data. Using the simple surflet combination feature F.21 the TPR of the support vector machine drops to 0.21. In Figure 4.10 colored scans of the wooden toy firetruck from the evaluation set are shown. The three central colors are red, blue, and cyan which correspond to cones, planes, and spheres, respectively. There are no tori or ellipsoids which is correct for the chosen object. Due to the dominance of cones in the selected feature, segments that might be cylindrical are classified as cones. Without homogenizing the density the classifier favors cylinders over

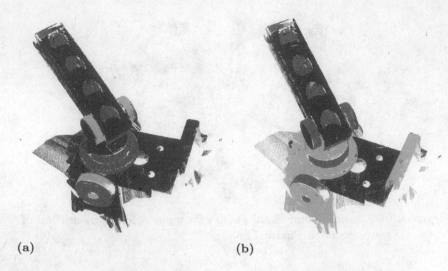

(a)   (b)

Figure 4.10: Scans of the wooden toy firetruck from the evaluation
set colored by primitive class using the simple surflet
combinations feature F.21 on homogenized data (a)
and non-homogenized data (a). The colors are: cones
(red), planes (blue), cylinders (magenta), spheres (cyan),
ellipsoids (green), tori (yellow).

cones which in some cases might be correct but often mistakes planes
for spheres or cones, see Figure 4.10b.

4.6 Conclusion

In this chapter a detailed evaluation of geometric features for primitive
classification in point clouds using support vector machines has been
presented. The approach has been compared to deep neural networks
using volumetric data for the classification of synthetic point clouds
and real 3d-scans. For training and evaluation, we computed a
synthetic data set, consisting of point cloud patches, by simulating
3d-scans of a laser line scanner. For comparison with deep neural

networks, an extensive voxel-based volumetric data set has been computed.

The geometric features are normal-, point-, and curvature-based. The discriminative power of single features, as well as feature combinations, has been evaluated. Feature F.21 achieved highest TPR among all presented features and feature combinations. It is surprising that this feature does not include any curvature information. Overall results of curvature-based features did not meet our expectations. Deep neural networks using volumetric data outperform support vector machines using geometric features on synthetic data. For segmented 3d-scans of real scenes and objects, the approach using geometric features is superior to neural networks. This result implies that it is harder for neural networks to generalize to real scans given a synthetic data set. We could show that homogenization improves the classification performance on real 3d-scans. Overall the performance of both methods is non-satisfactory for classification on the evaluation data set.

For future work, we intend to either create a large enough, labelled, data set of real 3d-scans or to enhance the method further by computing synthetic data to better match the properties of real scans. We plan to further investigate the intrinsic representation learned by neural networks. In [5] we trained denoising stacked autoencoders [102] on volumetric data of complete primitives. Visualizing the most prominent filter kernels, learned by the autoencoders, lead us to believe that symmetries play an essential role in primitive classification (see Figure 4.11 for examples). Understanding the more abstract concepts of feature correlations of deeper layers may enable the discovery of meaningful geometric representations. We also believe that voxel-based representations are inapt for many problems of geometric modeling. Learning methods that can be applied on arbitrary manifolds may be a viable alternative [7].

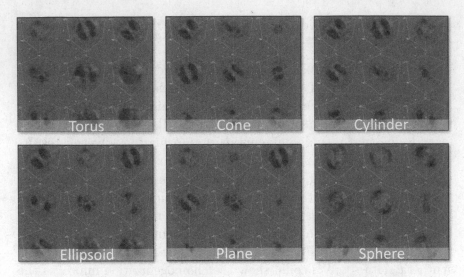

Figure 4.11: Visualization of 3d filter kernels with highest activation
per primitive class. The filters kernels are heatmap-colored
from low weight (blue) to high weight (red) [5].

5 CNN Texture Synthesis for High-Resolution Image Inpainting

In most modern manufacturing processes surface structure is stored separate from the basic geometry of the surface. A suitable counterpart from computer graphics are displacement maps which store a displacement on actual surface points to enrich models with detail. While displace-ment maps can have arbitrary depth resolution they are often represented as 8-bit grayscale images. This also applies to manufacturing processes where surface structure is represented in the form of, often tileable, 8-bit grayscale texure.

Image inpainting is the process of filling missing or corrupted regions in images based on surrounding image information so that the result looks visually plausible. To prevent the loss of structural information the resulting the grayscale images have to be of high-resolution. Recently machine learning techniques have been applied successfully to the problem of inpainting. In [72] Pathak et al. have shown that so-called context-encoders can fill missing regions in natural images. While being able to propagate global structure, the results lack details and are blurry. Because of this Yang et al. [110] propose a second network for detail transfer to the inpainting region after the content inpainting step. While this helps to compensate blurriness, it is not removed, and transitions in-between detail application patches are visible. This approach is still limited to lower resolution because of content inpainting and loss of global structure in the detail transfer process which renders it unsuitable for high-resolution textures.

© Springer Fachmedien Wiesbaden GmbH, part of Springer Nature 2020
P. Laube, *Machine Learning Methods for Reverse Engineering of Defective Structured Surfaces*, Schriftenreihe der Institute für Systemdynamik (ISD) und optische Systeme (IOS), https://doi.org/10.1007/978-3-658-29017-7_5

To resolve the issues mentioned above an inpainting approach that produces results that satisfy global structure and contain blurry-free details is proposed. The inpainting region is filled patch by patch using neural texture synthesis, which enables inpainting of high-resolution textures. We propose a convolutional neural network architecture that is able to shift focus from optimizing detail to global statistics and vice versa. Introduced in Section 2.1.3 spacial summary statistics computed from feature maps of deep convolutional neural networks are used for synthesis. For reference patch look-up we propose to use the same statistics that are used in patch synthesis.

Our method is not limited to the inpainting of grayscale heightmaps. The concept of neural texture synthesis has initially been introduced for color images. Inpainting of color images is the fundamental research topic in the inpainting community. For this reason, we will develop our approach on the basis of inpainting color images. In case of structured surfaces considered in the ToolRep project, a resolution of 2048×2048 pixels has shown to be sufficient to represent surface structure.

In the following, we will first give an overview of related works in Section 5.1. In Section 5.2 we present our inpainting approach, followed by the results and a discussion of the approach in Section 5.3.

5.1 Related Works

Most image inpainting approaches are based on sampling existing information surrounding the inpainting region, which is called exemplar-based inpainting [15, 106, 51, 24, 52, 23, 2]. Many exemplar-based methods approach inpainting in a copy-paste fashion. A well-known method is the work by Criminisi et al. [15] which applies this methodology but restricts the fill order to propagate linear structures first. This process has been further improved by Barnes et al. [2] who contribute a fast way for randomized correspondence search to find suitable patches. These approaches are dependent on patch-size and fail to satisfy global statistics that are not captured within a single patch. This inability to represent global statistics is a problem since texture is defined as a visual pattern modeled by a stationary stochastic process which includes global repetitions as well as local sub-patterns.

Recently deep learning approaches have been applied successfully to the problems of texture synthesis and image inpainting [60, 32, 47, 22]. First introduced by Gatys et al. in [31] texture synthesis by CNNs has shown to surpass well-known methods like [77] for many different examples. Wallis et al. [104] recently showed that artificial images produced from a parametric texture model closely match texture appearance for humans. Notably, the CNN texture model of [31] and the extension by Liu [65] can capture important aspects of material perception in humans. For many textures, the synthesis results are indiscriminate under foveal inspection. Other methods like the already mentioned methods by Phatak et al. [72] and Yang et al. [110] train auto-encoder like networks called context-encoders for inpainting. They use a combination of classic pixel-wise reconstruction loss as well as an adversarial loss term and a detail transfer network.

5.2 Patch-based texture synthesis for image inpainting

Applying the neural synthesis approach described in Section 2.1.3 we propose to synthesize texture Φ into some inpainting region Ω of an image Θ. Due to the high image resolution, we propose to inpaint Ω patch by patch. Figure 5.1a gives an overview of the different image regions in the inpainting process.

To authentically inpaint high-resolution textures one needs to satisfy not only the global appearance of texture but also the local details. In the leopard skin example of Figure 5.1a the global structure is represented in the form of the leopard patches while the detail structure is given by the fur itself.

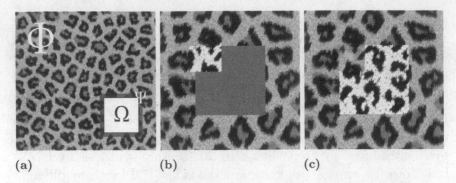

(a)   (b)   (c)

Figure 5.1: (a) Example image Θ of a leopard skin with inpainting region Ω, boundary Ψ and texture Φ. (b) After the inpainting of the first patch in the coarse inpainting step. (c) After the inpainting of the first patch in the detail application step

5.2.1 Synthesis Loss

Since the texture synthesis approach by Gatys et al. [31] fails to sufficiently synthesize long linear structures we include an additional loss term. Berger and Memisevic [4] introduce a loss term that incorporates the cross-correlation of spatially transformed feature maps. They achieve long-range consistency by computing the Gramian of some feature map $\mathbf{F}^l$ of layer l and a translated version of the feature map $T(\mathbf{F}^l)$ where $T(\cdot)$ is the translation operator. This translation is achieved by dropping either rows or columns of the respective feature map. This way the Gramian can be computed using the original feature map at some location $k = (x, y)$ and a shifted location $T_{x,+\delta}(k) = (x + \delta, y)$ or $T_{y,+\delta}(k) = (x, y + \delta)$. Like in Section 2.1.3 we define $\mathbf{F}^l_{ik}$ and $\mathbf{F}^l_{jk}$ to be the flattened feature maps i and j of the layer l, and k is the index of elements of the resulting feature map vector. The horizontally translated Gramian becomes

$$G^l_{x,\delta,ij} = \sum_k T_{x,+\delta}(F^l_{ik}) T_{x,-\delta}(F^l_{jk}), \qquad (5.1)$$

and $G^l_{y,\delta,ij}$ analogous. The cross-correlation loss $\mathcal{L}_{cc}$ for an arbitrary shift δ is defined as

$$\mathcal{L}_{cc}(\mathbf{x}, \widehat{\mathbf{x}}) = \sum_{l,i,j} \frac{(\mathbf{G}^l_{x,\delta,ij} - \widehat{\mathbf{G}}^l_{x,\delta,ij})^2 + (\mathbf{G}^l_{y,\delta,ij} - \widehat{\mathbf{G}}^l_{y,\delta,ij})^2}{4N_l^2 M_l^2}.$$

The combined loss together with (2.18) is then defined as

$$\mathcal{L}_{s,cc}(\mathbf{x}, \widehat{\mathbf{x}}) = w_s \mathcal{L}_s + w_{cc} \mathcal{L}_{cc},$$

with weight factors w_s and w_{cc} and $\mathcal{L}_s$ as defined in (2.18).

To achieve a smooth transition between texture Φ and inpainting region Ω the boundary region Ψ is included in the synthesis patch $\widehat{\mathbf{x}}$. By allowing Ψ in $\widehat{\mathbf{x}}$ and keeping Ψ almost steady throughout optimization the resulting synthetic texture has to satisfy the texture statistics while maintaining boundary Ψ. To limit the change of Ψ in the optimization an additional boundary-loss term is added.

We define the boundary loss as

$$\mathcal{L}_b(\mathbf{x}, \widehat{\mathbf{x}}) = \frac{1}{P} \sum (\mathbf{mx} - \mathbf{m}\widehat{\mathbf{x}})^2, \qquad (5.2)$$

where the binary mask $\mathbf{m}$ equals 0, if $\mathbf{m}_i \in \Omega$, and 1 otherwise.

5.2.2 Two-Scale Resolution Synthesis

To be able to process high-resolution textures globally as well as locally a two-resolution neural network architecture is applied. Figure 5.2 gives an overview of our network architecture and the different entities involved in the inpainting process. The top left shows the image for inpainting together with essential regions. The *reference CNNs* compute the global and detail Gramians. The *synthesis CNN* shows the architecture of the network for inpainting together with the different loss terms that lead to the global loss $\mathcal{L}$. At the bottom, a legend of the different image elements is shown. The synthesis

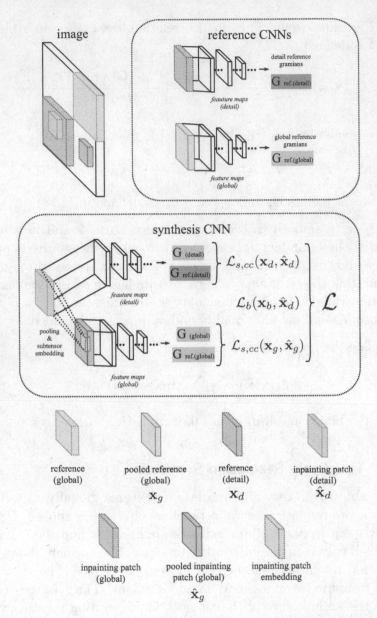

Figure 5.2: Schematic overview of the different entities and neural networks involved in the inpainting approach.

CNN consists of the *global branch* with focus on optimization for global statistics and the *detail branch* which optimizes to satisfy detail statistics. In Figure 5.2 the global branch is the bottom branch of the synthesis CNN, and the detail branch is at the top. Each branch consists of a VGG19 network as outlined in Section 2.1.3. Each VGG19-instance has been pre-trained on the ImageNet [20] dataset for classification.

The input to the detail branch is defined as $\widehat{\mathbf{x}}_d$, and the input to the global branch as $\widehat{\mathbf{x}}_g$. Input $\widehat{\mathbf{x}}_d$ is the actual patch (green in Figure 5.2), containing part of Ω, that will be optimized in native resolution. Input $\widehat{\mathbf{x}}_g$ is initialized with a Q-times average-pooled cutout of Θ so that this cutout fully contains Ω and boundary Ψ. Interconnection of the two branches is achieved by subtensor-embedding where $\widehat{\mathbf{x}}_d$ is inserted into $\widehat{\mathbf{x}}_g$. To match the resolution for embedding, Q average-pooling-layers are placed in-between $\widehat{\mathbf{x}}_d$ and $\widehat{\mathbf{x}}_g$. These pooling layers have a window size of 2x2. While $\widehat{\mathbf{x}}_d$ grants a detailed close-up on the inpainting region, $\widehat{\mathbf{x}}_g$ represents a larger region of the texture containing the inpainting region and its surroundings. Only $\widehat{\mathbf{x}}_d$ will be updated in the optimization process. By embedding $\widehat{\mathbf{x}}_d$ in $\widehat{\mathbf{x}}_g$ through the network architecture $\widehat{\mathbf{x}}_g$ is updated automatically in the forward propagation step.

Depending on the size of Ω, Q needs to be adjusted as a parameter before inpainting. As a result, the network topology needs to be adapted. Since this change does not introduce new weights, but only pooling layers, it is easily adaptable without training for different sizes of Ω or different resolutions. The inpainting patch $\widehat{\mathbf{x}}_d$ is shifted along the inpainting region so that a new patch position overlaps with the old one and with Ψ.

At the beginning of the inpainting process global reference $\mathbf{x}_g$ needs to be initialized. The detail reference $\mathbf{x}_d$ needs to be reinitialized with a new reference patch from Φ after each shift of the inpainting patch $\widehat{\mathbf{x}}_d$ to a new location. Reference patch look-up is described in Section 5.2.3.

The combined loss over both branches together with the boundary loss becomes

$$\mathcal{L} = w_d \mathcal{L}_{s,cc}(\mathbf{x}_d, \widehat{\mathbf{x}}_d) + w_g \mathcal{L}_{s,cc}(\mathbf{x}_g, \widehat{\mathbf{x}}_g) + w_b \mathcal{L}_b(\mathbf{x}_b, \widehat{\mathbf{x}}_d),$$

where w_d, w_g, and w_b are weight terms. $\mathbf{x}_b$ is initialized with $\widehat{\mathbf{x}}_d$ before optimization and does change for each new position of $\widehat{\mathbf{x}}_d$.

5.2.3 Reference Patch Look-up

To find suitable reference patches $\mathbf{x}_d$, and $\mathbf{x}_g$ in Φ we propose to use Gramians as defined in (2.17). After the initialization $\widehat{\mathbf{x}}_d$ contains a cut-out of Θ. This cut-out not only contains the inpainting region Ω but also parts of the boundary region Ψ. Initial $\widehat{\mathbf{x}}_g$ completely contains Ψ and Ω. The size of the boundary Ψ depends on the size of the inpainting region Ω and the number of poolings Q.

For synthesis, a reference patch that best matches the current $\widehat{\mathbf{x}}_d$ respectively the initial $\widehat{\mathbf{x}}_g$ is needed. The region for the look-up is restricted to Φ. Most approaches use the mean squared error to find the closest match. Instead of a mean squared error, we propose to use the distance of Gramians of the same layer as a measure of similarity. Because values inside inpainting region Ω are unknown we propose masking Ω for each feature map to remove Ω related correlations from any of the resulting Gramians. Using these feature map masks, the Gramian of $\widehat{\mathbf{x}}_d$ and $\widehat{\mathbf{x}}_g$ can be computed independently of Ω.

Up to some layer l, the network input will pass through different pooling and convolutional layers. Feature map masks need to be adapted to compensate for these operations. In a first step the initial, binary mask $\mathbf{m}$ from Eq. (5.2), needs to be pooled similar to the input. This is achieved by applying each max-pooling step of VGG-19 that has been applied up to layer l to the mask $\mathbf{m}^l$. Each $\mathbf{m}^l$ is responsible for masking feature maps $\mathbf{F}^l$. In a second step, one needs to account for the propagation of Ω into Ψ by convolutions. Convolutional filters of VGG-19 are of size 3x3. Therefore the inpainting region Ω

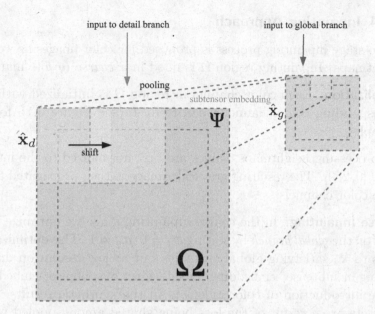

Figure 5.3: Overview of the subtensor embedding process. First $\widehat{\mathbf{x}}_g$ is initialized with a pooled cut-out of image Φ that fully contains Ψ and Ω. After each optimization step of $\widehat{\mathbf{x}}_d$, $\widehat{\mathbf{x}}_g$ is updated with pooled $\widehat{\mathbf{x}}_d$ by subtensor embedding. The colors of the different regions are the same as in Figure 5.2.

propagates into Ψ one pixel at a time for each convolutional layer. This propagation can be compared to binary-erosion from image processing. Simply discarding those pixels by setting them to zero in $\mathbf{m}^l$ for each convolutional layer is too restrictive and would lead to masks with all values zero in late layers. Expansion of Ω by a smaller, individual number of pixels e^l for each layer is applied. In our experiments, this expansion has proven to be sufficient for compensation.

Taking these considerations into account we define patch distance as

$$\Delta_{\mathbf{G}}(\mathbf{x}, \widehat{\mathbf{x}}) = \sum_{l,i,j} (\sum_k \mathbf{m}^l \mathbf{F}_{ik}^l \mathbf{F}_{jk}^l - \sum_k \mathbf{m}^l \widehat{\mathbf{F}}_{ik}^l \widehat{\mathbf{F}}_{jk}^l)^2.$$

5.2.4 Inpainting Approach

A two-stage inpainting process is proposed for color images as well as heightmaps. Inpainting region Ω is filled in a *coarse to fine* manner:

Initialization: Each color channel in region Ω is initialized with the corresponding color channel mean from Φ (see Figure 5.1b for an example).

For processing heightmaps, values are first normalized to the integer range $[0, 255]$. The resulting grayscale image is then propagated to all three color channels.

Coarse Inpainting: In the coarse inpainting stage we optimize with focus on the *global branch* by setting $w_d = 0$, $w_g = 1$. This optimization leads to $\widehat{\mathbf{x}}_d$ satisfying global statistics but at low resolution due to pooling in subtensor-embedding. A side-effect of subtensor-embedding is the introduction of color artefacts in the synthetic results. The artefacts are a result of the loss being shared among pooled pixels as can be seen in Figure 5.1c. We eliminate these color artifacts by converting Ω to grayscale (see Figure 5.1c) with RGB weights $r = 0.212$, $g = 0.7154$, $b = 0.0721$. For heightmap input, this conversion to grayscale is skipped. Since color channels of heightmap input will diverge in the optimization of synthetic texture, they are averaged after each optimization step. Each color channel is then reinitialized with this average.

After inpainting region, Ω has been filled with the coarse texture we proceed with detail application.

Detail Application: In the detail application stage we set $w_d = 1$ and w_g to a value in the range of $[0.01, 0.1]$. The weights ensure focus on the optimization for detail statistics through the *detail branch* while constraining the optimization to also maintain global texture statistics (see Figure 5.1c for an example). Again, color channels are averaged and propagated back to all color channels for heightmap input.

Fill Order: A critical consideration in well-known patch-based inpainting methods like [15] is the order in which to fill Ω. For our approach this order is not essential as long as the first patch overlaps with Ψ. We also make sure that consecutive patches overlap. Overlapping a patch by $\frac{1}{4}$ of its size with surrounding texture has proven to be sufficient. To ensure a smooth transition in-between patches, we apply image quilting [23] in overlapping regions. Region Ω is inpainted in a top to bottom, left to right fashion, both for the coarse and the detail step.

As a results of our experiments we set $w_s = 1e6$ and $w_{cc} = 1e7$ for inpainting of 8-bit color images as well as heightmaps. Choosing w_b in the range $[10, 50]$ has shown to be sufficient. The significant difference between Gramian-based loss weights and weights related to loss in pixel space results from different value ranges. We optimize $\widehat{\mathbf{x}}_d$ by applying the L-BFGS-B [115] algorithm like proposed in [32].

5.3 Results

In this Section, we present the results of the proposed inpainting approach. Inpainting results are shown and compared on the basis of high-resolution color textures. Also, results for inpainting high-resolution heightmaps are presented. All textures have a resolution of 2048x2048 px while the inpainting region Ω is of size 512x512 px. For initialization of the reference and synthesis CNN a pre-trained VGG-19 network, provided by [31], is used. Parameters of the network are transferred to a theano-based [96] implementation. The network input size, respectively the inpainting patch size, for all networks is 256x256 px. For definition of $\mathcal{L}$ we use Gramians of layers *conv1_1*, *pool1*, *pool2*, *pool3* and *pool4* leading to $L = 5$. In our experiments layer *pool4* sometimes led to distortions in the patch overlap regions. For inpainting results with strong distortions we propose to replace *pool4* with *pool5*. For patch distance computation we define pixel expansions $\mathbf{e} = (1, 1, 2, 3, 2)$, and for shift δ of translated Gramians $\mathbf{G}_{x,\delta}^l$ and $\mathbf{G}_{y,\delta}^l$ we define $\boldsymbol{\delta} = (6, 6, 5, 4, 3)$. For our examples $Q = 2$

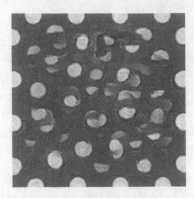

Figure 5.4: Inpainting results of the context-encoder-based method
by Yang et al. [110]. The approach is unable to fill
inpainting region Ω with meaningful texture. The provided
implementation by Yang et al. was limited to a maximum
resolution of 512x512 px.

pooling layers suffice to ensure that Ω as well as Ψ fit into $\widehat{\mathbf{x}}_g$. To
find suitable reference patches $\mathbf{x}_d$ and $\mathbf{x}_g$ a step size of 64 px is used
for look-up. Due to the Gramian-based inpainting strategy reference
patches do not need to exactly match the current $\mathbf{x}_d$ and $\mathbf{x}_g$ regarding
MSE. Therefore a more significant step size suffices which speeds
up reference patch look-up. For periodic textures without much
stochasticity using the MSE for patch look-up will be superior to the
Gramian-based patch look-up. Remarks regarding the runtime can
be found in appendix A.2.

We compare our approach for patch-based inpainting by texture
synthesis (PITS) to the exemplar-based method of Criminisi et al.
[15] (EI) as well as the Content-Aware Fill method of Photoshop CC17
(PSCC17) which is a combination of methods [2] and [106]. We do not
compare to method [110] using context-encoders since the approach
seems to be unable to sufficiently inpaint texture and is further unable
to process images larger 512x512 px (see Figure 5.4). We compare
results for each example of original texture and inpainted Ω. Since
distance measures like the structural dissimilarity measure [105], or

the mean squared error are inadequate for comparison inpainting results are evaluated visually. We also present a visual evaluation in Fig. 5.5.

For the first exemple texture *Dots* (Figure 5.5a) PITS achieves best visual quality. While EI fails to inpaint Ω reasonably, the result of PSCC17 looks visually coherent but fails to reconstruct dots as perfect circles. PITS manages to produce visually appealing structure globally as well as locally. For the *Bricks* example in Fig. 5.5b it is clearly visible that EI again fails to provide a visually likely result. Regarding structural continuity PITS produces better-looking results than PSCC17. When enlarged one can see that for the *Bricks* example our method leads to smaller changes in the color palette in-between inpainted patches which is a result of the color variation between reference patches. For the *Wood* texture PITS inpaints Ω reasonably while EI creates a visually uncompelling result. Result by PSCC17 look reasonable, but knotholes seem to have been directly copied over from Φ to Ω. In figures 5.7 and 5.8 we present additional results of PITS, PSCC17, and EI applied to high-resolution textures. The presented heightmaps are representative of the heightmaps used for the production of functional surfaces. Compared to results of PSCC17, results of the PITS method are visually unsatisfactory. Results of PITS contain noise, are distorted and show artifacts in regions where patches connect. These problems will be discussed in the following section.

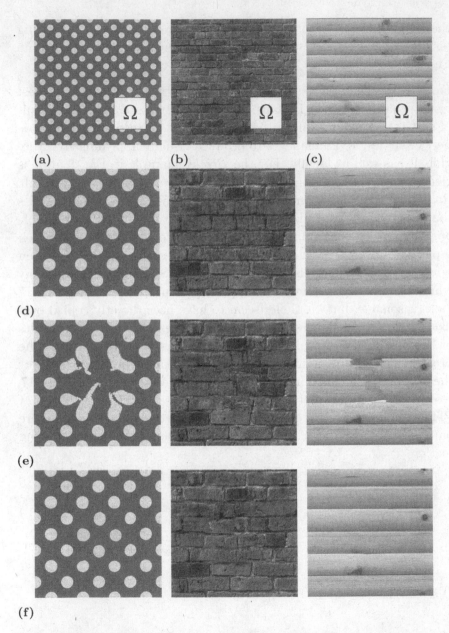

Figure 5.5: Original with inpainting region Ω (a)-(c). Close-up on inpainting results of PITS (c), EI (d), and PSCC17 (e).

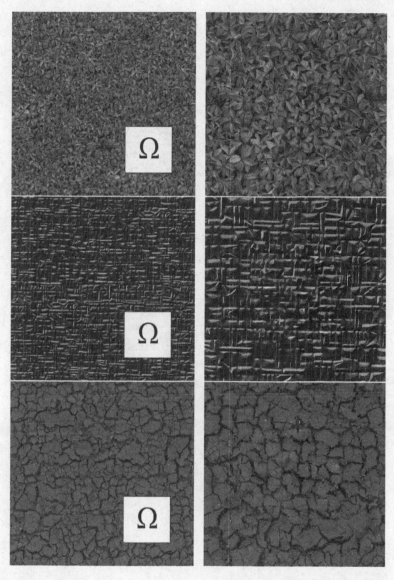

Figure 5.6: Examples of inpainting results produced by PITS with the
original texture on the left and a close-up on the inpainted
region on the right.

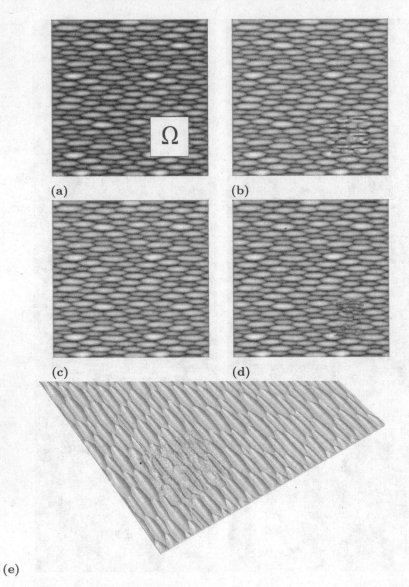

Figure 5.7: Result for inpainting of a hammer stroke texture heightmap
(a). Result of the PITS method (b), of PSCC17 (c), and of
EI (d). The reconstructed triangular mesh of (b) is shown in
(e).

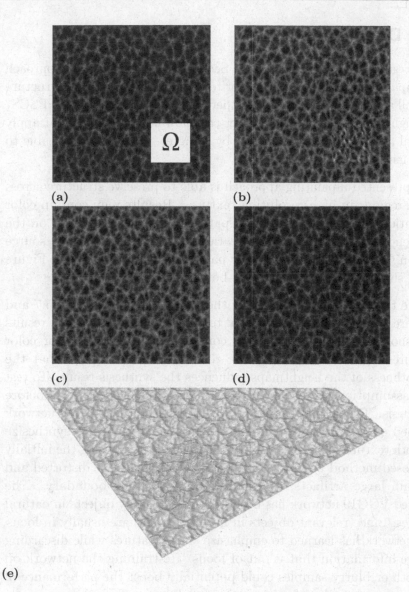

Figure 5.8: Result for inpainting of a leather grain texture heightmap
(a). Result of the PITS method (b), of PSCC17 (c), and of
EI (d). The reconstructed triangular mesh of (b) is shown in
(e).

5.4 Discussion

The results for color images from Section 5.3 show that our approach for inpainting high-resolution color texture can satisfy global structure as well as details. It is able to outperform methods like EI and PSCS7 for high-resolution textures. Other comparable approaches that apply neural networks like the method by Yang et al. [110] are not able to sufficiently inpaint textures.

The presented inpainting approach is able to preserve structure across large regions in high-resolution textures. Results may contain color deviations in-between inpainted patches depending strongly on the reference patch for detail application. Especially when the source region Φ has a very diverse color palette, like the example in Figure 5.5b, shifts in color are visible in-between inpainting patches.

While the inpainted textures for the heightmaps in Figures 5.7 and 5.8, are comparable to the original texture to a certain extend, results fall short of expectations when compared to the results for color textures. On the basis of these results one can assume that the smoothness of the heightmaps influences the synthesis result. To test this assumption Gaussian blur has been applied to textures before synthesis. The results in Figure 5.10 show that the VGG19 network trained on the ImageNet data set is unable to correctly synthesize smooth texture. The textures have been synthesized using the initially proposed method by Gatys et al. [31]. The results are distorted and contain large artifacts, especially near the image boundary. The applied VGG19 network has been trained to classify objects in natural images. Since relevant objects in natural images are usually in focus, the network has learned to emphasize sharp features while discarding image information that is out of focus. Re-training the network on smooth or blurry samples could potentially boost the performance of the proposed method for inpainting heightmaps.

Using the difference of masked Gramians as a patch distance measure helps to reduce the number of reference patches that need to be

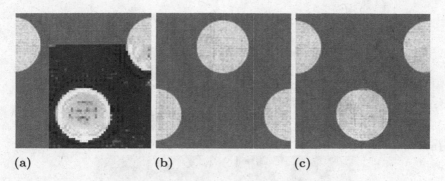

(a)   (b)   (c)

Figure 5.9: (a) Inpainting patch $\widehat{\mathbf{x}}_d$. (b) Closest reference patch from Φ.
c Inpainting result.

compared in the reference look-up step. This is a consequence of reference textures $\mathbf{x}_d$ or $\mathbf{x}_g$ not needing to exactly match Ψ inside $\widehat{\mathbf{x}}_d$ or $\widehat{\mathbf{x}}_g$ in terms of mean squared error. Due to the averaging of feature information inside Gramians, global spatial information is lost. Loss of spatial information enables the Gramian to represent texture invariant to rotation and translation to some degree. Figure 5.9 shows an example where a reference texture with large mean squared error to the inpainting patch leads to a sound inpainting result.

A significant strength of the dual resolution approach is the correct synthesis of global structure while also maintaining details. When choosing w_d and w_g one needs to be aware of the trade-off introduced. While higher w_g ensures persistence of global statistics, it also introduces artefacts as a result of pooling $\widehat{\mathbf{x}}_d$ before subtensor embedding and vice versa. Higher w_d lays more considerable emphasis on details while possibly violating global structure. The number of poolings Q further influences this trade-off.

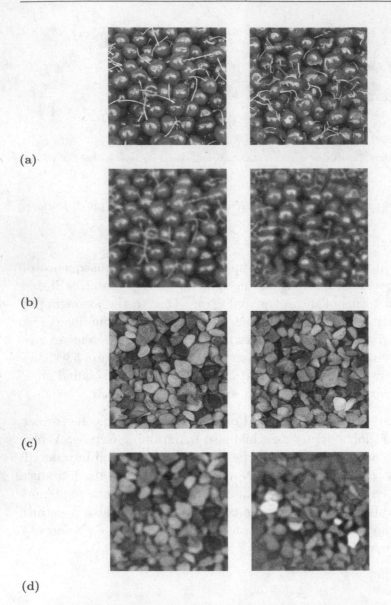

(a)

(b)

(c)

(d)

Figure 5.10: (a) and (c) show synthesis results (right) of the original
textures (left). (b) and (d) show the same textures after
the application of Gaussian blur (left) and the resulting
synthesis (right).

5.5 Conclusion

In Chapter 5, we presented a new method based on CNN texture synthesis for inpainting. The approach can inpaint missing regions of large-scale, high-resolution textures. The texture is processed by dual-resolution synthesis. We apply a neural network that consists of a global and a local branch for the optimization of global respectively local texture statistics. This way our approach avoids the problems of missing details from which previous CNN approaches suffered, while plausibly continuing global image structure. In principle, this network architecture can be extended to include a hierarchy of more than two interacting scales. We evaluated our method on the basis of color textures as well as height maps.

While the approach can inpaint color textures reasonably, it has problems synthesizing smooth heightmaps of surface structure. Retraining the VGG19 model on smooth or blurry data to improve the inpainting results of heightmaps is a significant focus of our ongoing research. The design of a multi-resolution architecture, including more than two resolutions, could be an exciting line of research that we plan to pursue in the future.

6 Concluding Remarks & Outlook

In this thesis, I presented novel machine learning approaches for key problems of the reverse engineering process of structured surfaces. The developed methods aim to further automate the steps of reverse engineering further. A particular emphasis is laid on the reconstruction of the surface's base geometry as well as surface structure which is mandatory for many functional surfaces. In the case of the ToolRep project the scanned functional surfaces also contain local surface defects. Three key problems of reverse engineering, the automation of which will further contribute to process automation, have been identified. These key problems are reflected by chapters 3 to 5.

Methods for CAD reconstruction of scanned functional surfaces need to cope with freeform surface approximation as well as geometric primitive fitting which includes primitive recognition. An important, unsolved problem of freeform surface approximation is the computation of parametrizations. In Chapter 3, parametrizations for curve and surface approximation are predicted. While Section 3.2 covered the application of SVMs for knot vector computation, Section 3.3 introduced neural networks for the computation of point parametrizations as well as knot vectors. It could be shown that both methods can produce tight approximations, and can outperform state of the art methods for parametrization. While the scope of Sections 3.2 and 3.3 is limited to the approximation of B-spline curves, Section 3.4 evaluated both methods in the context of T-spline surface approximation. Again both methods result in tight approximations with an advantage for the SVM-based method. Since parts of functional surfaces can often be represented by geometric primitives, detection of such primitives is

© Springer Fachmedien Wiesbaden GmbH, part of Springer Nature 2020
P. Laube, *Machine Learning Methods for Reverse Engineering of Defective Structured Surfaces*, Schriftenreihe der Institute für Systemdynamik (ISD) und optische Systeme (IOS), https://doi.org/10.1007/978-3-658-29017-7_6

crucial to the fitting procedure. In Chapter 4 an extensive evaluation of geometric features for primitive classification is presented. SVMs have been trained to classify six primitive classes. Classification performance has been compared to neural networks using a voxelized point cloud representation. While the classification by neural networks results in a higher TPR for synthetic data, SVMs perform better on real scans. Overall, both methods fall short of expectations on real scans. Since surface structure of most functional surfaces is represented by high-resolution grayscale texture, fixing defects in surface structure is regarded as an image inpainting problem. In Chapter 5 a novel method, applying neural texture synthesis for high-resolution image inpainting, is presented. The two-scale resolution synthesis can convincingly inpaint large regions in color images. While the approach successfully inpaints color images it is not able to sustain high-quality results when applied to heightmaps. This is a result of the inability of the VGG19 network to represent smooth texture.

The evaluation of the presented methods indicates that machine learning will be an important part of geometric modeling and reverse engineering in the future. The quantity of current research on the topic proves the attractiveness and efficiency of the methods. While classical algorithms of image processing have already been influenced or replaced more and more by machine learning methods, their application to problems of geometry processing is still small. In Section 3.2 and Chapter 4 geometric features are used to represent point data while in Section 3.3 neural networks are applied without an intermediate representation. In geometry processing, it has been common practice to extract features using expert knowledge. I am convinced that future algorithms will independently find suitable representations. Understanding these recovered representations can open new perspectives and views on known problems.

While the performance of machine learning methods is impressive, they are subject to specific critical requirements for success. A significant hurdle is the large amount of training data required. While numerous high-quality databases with labelled image data exist, the amount of existing labelled geometric data is small. In the respective chapters, we have tried to compensate for this absence of data by creating synthetic data but reached certain limitations. Models which, given synthetic data, generalize well to real-world data are challenging to train. Various tests with methods for unsupervised learning, which are not presented in this thesis, often stay behind supervised methods in terms of success rate. The creation of real-world 3d data sets is, in contrast to image processing, extremely complex. The generation and manual labelling of scans is a complex task which needs to be done by experts. I also do believe, that defining the problem at hand in an optimizable fashion, as we did in Section 3.3 by introducing B-spline approximation into the neural network architecture, the dependency on correctly labelled real-world data can further be loosened. Another challenge is generalization across multiple resolutions. In Chapter 5 a two-scale resolution approach has been introduced to represent global as well as local texture features. While resolution is well defined in image space, it is not clear how to apply such a concept to geometry. The problem becomes clear by an examination of the primitive classification problem. Locally, point clouds may be represented by small planes which does not reflect an objects inherent geometry. Finding the proper level of resolution for approximation is crucial to success. In Section 3.3 we approached this problem using segmentation, subsampling, and supersampling of the point sets.

The processing of detailed surface structures in image space allows the use of algorithms originating from image processing. However, Chapter 5 shows the strong relation of training data and successful generalization. While the neural network successfully inpaints color images, results on heightmaps are non-satisfactory. For future research, I suggest to skip the separation of surface structure and base geometry and inpaint the surface directly. I regard recent attempts in manifold learning as an exciting new field of research. Defining significant operations like convolutions on a graph, render an intermediate representation like pixels or voxels unnecessary. First achievements in surface synthesis, especially in alliance with unsupervised methods like the work by Litany et al. [64], confirm this assumption. In this thesis I regard defects of the base geometry as repaired after approximation and fitting. Complex geometries and more substantial defects may require the application of surface synthesis. The efficiency of neural networks for image processing has led to the development of specialized frameworks and hardware for efficient computation. These computations strongly rely on the existence of a regular grid which is not given for geometric data. Neural networks that can be applied to graphs may be better suited for the development of frameworks and the computation on GPU hardware.

While the three isolated key-problems of reverse engineering of defective structured surfaces are pillars of the process, there exist many more steps that may benefit from machine learning. The above-mentioned point cloud segmentation is a problem of particular importance. Like presented in [80], neural networks trained for classification can also be used to segment scenes. For the future, I would like to apply the presented classification methods to segmentation. I further am currently working on defect detection in surface structure heightmaps using the neural network representations of Section 5. Transferring this concept to surface analysis is another area of research I would like to approach in the future.

Bibliography

[1] Georg Arbeiter, Steffen Fuchs, Richard Bormann, Jan Fischer, and Alexander Verl. Evaluation of 3d feature descriptors for classification of surface geometries in point clouds. In *IEEE/RSJ International Conference on Intelligent Robots and Systems*, pages 1644–1650, 2012.

[2] Connelly Barnes, Eli Shechtman, Adam Finkelstein, and Dan B Goldman. Patchmatch: A randomized correspondence algorithm for structural image editing. *ACM Trans. Graph.*, 28(3):24–1, 2009.

[3] Yoshua Bengio, Aaron Courville, and Pascal Vincent. Representation learning: A review and new perspectives. *IEEE transactions on pattern analysis and machine intelligence*, 35(8):1798–1828, 2013.

[4] Guillaume Berger and Roland Memisevic. Incorporating long-range consistency in cnn-based texture generation. *arXiv preprint arXiv:1606.01286*, 2016.

[5] Merlin Blume. 3d primitive classification using stacked autoencoders. Technical report, University of Konstanz, 2015.

[6] Léon Bottou and Chih-Jen Lin. Support vector machine solvers. *Large scale kernel machines*, 3(1):301–320, 2007.

[7] Michael M Bronstein, Joan Bruna, Yann LeCun, Arthur Szlam, and Pierre Vandergheynst. Geometric deep learning: going beyond euclidean data. *IEEE Signal Processing Magazine*, 34(4):18–42, 2017.

© Springer Fachmedien Wiesbaden GmbH, part of Springer Nature 2020
P. Laube, *Machine Learning Methods for Reverse Engineering of Defective Structured Surfaces*, Schriftenreihe der Institute für Systemdynamik (ISD) und optische Systeme (IOS), https://doi.org/10.1007/978-3-658-29017-7

[8] M Caputo, K Denkèr, M Franz, P Laube, and G Umlauf. Learning geometric primitives in point clouds. In *Eurographics Symposium on Geometry Processing SGP (2014)*. Eurographics Association, 2014.

[9] Manuel Caputo, Klaus Denker, Mathias O Franz, Pascal Laube, and Georg Umlauf. Support vector machines for classification of geometric primitives in point clouds. In *Curves and Surfaces*, pages 80–95. Springer, 2014.

[10] F. Cazals and M. Pouget. Estimating differential quantities using polynomial fitting of osculating jets. *Computer Aided Geometric Design*, 22(2):121–146, 2005.

[11] O. Chapelle, P. Haffner, and V. Vapnik. Support vector machines for histogram-based image classification. *IEEE Trans. on Neural Networks*, 10(5):1055–1064, 1999.

[12] Xiao-Diao Chen, Weiyin Ma, and Jean-Claude Paul. Cubic b-spline curve approximation by curve unclamping. *Computer-Aided Design*, 42(6):523–534, 2010.

[13] Xiaozhi Chen, Kaustav Kundu, Ziyu Zhang, Huimin Ma, Sanja Fidler, and Raquel Urtasun. Monocular 3d object detection for autonomous driving. In *Proceedings of the IEEE Conference on Computer Vision and Pattern Recognition*, pages 2147–2156, 2016.

[14] Xiaozhi Chen, Huimin Ma, Ji Wan, Bo Li, and Tian Xia. Multiview 3d object detection network for autonomous driving. In *IEEE CVPR*, volume 1, page 3, 2017.

[15] Antonio Criminisi, Patrick Pérez, and Kentaro Toyama. Region filling and object removal by exemplar-based image inpainting. *IEEE Transactions on image processing*, 13(9):1200–1212, 2004.

[16] Nello Cristianini and John Shawe-Taylor. *An introduction to support vector machines and other kernel-based learning methods*. Cambridge University Press, 2000.

[17] Malvin Danhof, Tarek Schneider, Pascal Laube, and Georg Umlauf. A virtual-reality 3d-laser-scan simulation. pages 68–73, 2015.

[18] Paul De Casteljau. Outillages méthodes calcul. *Andr e Citro en Automobiles SA, Paris*, 1959.

[19] Paul De Casteljau. Courbes et surfaces à pôles. *André Citroën, Automobiles SA, Paris*, 1963.

[20] Jia Deng, Wei Dong, Richard Socher, Li-Jia Li, Kai Li, and Li Fei-Fei. Imagenet: A large-scale hierarchical image database. In *Computer Vision and Pattern Recognition, 2009. CVPR 2009. IEEE Conference on*, pages 248–255. Ieee, 2009.

[21] K. Denker, D. Hagel, J. Raible, G. Umlauf, and B. Hamann. On-line reconstruction of CAD geometry. In *Intl. Conf. on 3D Vision*, pages 151–158, 2013.

[22] Alexey Dosovitskiy and Thomas Brox. Generating images with perceptual similarity metrics based on deep networks. In *Advances in Neural Information Processing Systems*, pages 658–666, 2016.

[23] Alexei A Efros and William T Freeman. Image quilting for texture synthesis and transfer. In *Proceedings of the 28th annual conference on Computer graphics and interactive techniques*, pages 341–346. ACM, 2001.

[24] Alexei A Efros and Thomas K Leung. Texture synthesis by non-parametric sampling. In *Computer Vision, 1999. The Proceedings of the Seventh IEEE International Conference on*, volume 2, pages 1033–1038. IEEE, 1999.

[25] Megumi Endoh, Tomohiro Yanagimachi, and Ryutarou Ohbuchi. Efficient manifold learning for 3d model retrieval by using clustering-based training sample reduction. In *ICASSP*, pages 2345–2348, 2012.

[26] Gerald E Farin. *Curves and surfaces for CAGD: a practical guide*. Morgan Kaufmann, 2002.

[27] Rida T Farouki. Optimal parameterizations. *Computer Aided Geometric Design*, 14(2):153–168, 1997.

[28] Manuel Fernández-Delgado, Eva Cernadas, Senén Barro, and Dinani Amorim. Do we need hundreds of classifiers to solve real world classification problems. *Journal of Machine Learning Research*, 15(1):3133–3181, 2014.

[29] Akemi Gálvez and Andrés Iglesias. Firefly algorithm for explicit b-spline curve fitting to data points. *Mathematical Problems in Engineering*, 2013, 2013.

[30] Akemi Gálvez, Andrés Iglesias, Andreina Avila, César Otero, Rubén Arias, and Cristina Manchado. Elitist clonal selection algorithm for optimal choice of free knots in b-spline data fitting. *Applied Soft Computing*, 26:90–106, 2015.

[31] Leon Gatys, Alexander S Ecker, and Matthias Bethge. Texture synthesis using convolutional neural networks. In *Advances in Neural Information Processing Systems*, pages 262–270, 2015.

[32] Leon A Gatys, Alexander S Ecker, and Matthias Bethge. Image style transfer using convolutional neural networks. In *Proceedings of the IEEE Conference on Computer Vision and Pattern Recognition*, pages 2414–2423, 2016.

[33] Xavier Glorot, Antoine Bordes, and Yoshua Bengio. Deep sparse rectifier neural networks. In *Proceedings of the Fourteenth International Conference on Artificial Intelligence and Statistics*, pages 315–323, 2011.

[34] Rony Goldenthal and Michel Bercovier. Spline curve approximation and design by optimal control over the knots. In *Geometric Modelling*, pages 53–64. Springer, 2004.

[35] Ian Goodfellow, Jean Pouget-Abadie, Mehdi Mirza, Bing Xu, David Warde-Farley, Sherjil Ozair, Aaron Courville, and Yoshua Bengio. Generative adversarial nets. In *Advances in neural information processing systems*, pages 2672–2680, 2014.

[36] Michael Grunwald, Matthias Hermann, Fabian Freiberg, Pascal Laube, and Matthias O Franz. Optical surface inspection: A novelty detection approach based on cnn-encoded texture features. In *Applications of Digital Image Processing XLI*, volume 10752, page 107521E. International Society for Optics and Photonics, 2018.

[37] Michael Grunwald, Pascal Laube, Martin Schall, Georg Umlauf, and Matthias O Franz. Radiometric calibration of digital cameras using neural networks. In *Optics and Photonics for Information Processing XI*, volume 10395, page 1039505. International Society for Optics and Photonics, 2017.

[38] Michael Grunwald, Jens Müller, Martin Schall, P Laube, G Umlauf, and MO Franz. Pixel-wise hybrid image registration on wood decors. In *Symposium on Information and Communication Systems BW-CAR— SINCOM (2015)*, pages 24–29, 2015.

[39] Steve R Gunn et al. Support vector machines for classification and regression. *ISIS technical report*, 14:85–86, 1998.

[40] Igor Guskov, Wim Sweldens, and Peter Schröder. Multiresolution signal processing for meshes. In *Proceedings of the 26th annual conference on Computer graphics and interactive techniques*, pages 325–334. ACM Press/Addison-Wesley Publishing Co., 1999.

[41] Jun Han and Claudio Moraga. The influence of the sigmoid function parameters on the speed of backpropagation learning. In *International Workshop on Artificial Neural Networks*, pages 195–201. Springer, 1995.

[42] G. Hetzel, B. Leibe, P. Levi, and B. Schiele. 3d object recognition from range images using local feature histograms. In *CVPR*, pages 394–399, 2001.

[43] Geoffrey E Hinton, Nitish Srivastava, Alex Krizhevsky, Ilya Sutskever, and Ruslan R Salakhutdinov. Improving neural networks by preventing co-adaptation of feature detectors. *arXiv preprint arXiv:1207.0580*, 2012.

[44] Kurt Hornik, Maxwell Stinchcombe, and Halbert White. Multilayer feedforward networks are universal approximators. *Neural networks*, 2(5):359–366, 1989.

[45] Chih-Wei Hsu, Chih-Chung Chang, and Chih-Jen Lin. A practical guide to support vector classification. Technical report, Department of Computer Science, National Taiwan University, Taiwan, 2003.

[46] Sergey Ioffe and Christian Szegedy. Batch normalization: Accelerating deep network training by reducing internal covariate shift. *arXiv preprint arXiv:1502.03167*, 2015.

[47] Justin Johnson, Alexandre Alahi, and Li Fei-Fei. Perceptual losses for real-time style transfer and super-resolution. In *European Conference on Computer Vision*, pages 694–711. Springer, 2016.

[48] Diederik P Kingma and Jimmy Ba. Adam: A method for stochastic optimization. *arXiv preprint arXiv:1412.6980*, 2014.

[49] Stefan Knerr, Léon Personnaz, and Gérard Dreyfus. Single-layer learning revisited: a stepwise procedure for building and training a neural network. In *Neurocomputing*, pages 41–50. Springer, 1990.

[50] J.J. Koenderink and A.J. van Doorn. Surface shape and curvature scales. *Image Vision Comput.*, 10(8):557–565, 1992.

[51] Vivek Kwatra, Irfan Essa, Aaron Bobick, and Nipun Kwatra. Texture optimization for example-based synthesis. *ACM Transactions on Graphics (ToG)*, 24(3):795–802, 2005.

[52] Vivek Kwatra, Arno Schödl, Irfan Essa, Greg Turk, and Aaron Bobick. Graphcut textures: image and video synthesis using graph cuts. In *ACM Transactions on Graphics (ToG)*, volume 22, pages 277–286. ACM, 2003.

[53] Pascal Laube, Matthias O Franz, and Georg Umlauf. Deep learning parametrization for b-spline curve approximation. In *2018 International Conference on 3D Vision (3DV)*, pages 691–699. IEEE, 2018.

[54] Pascal Laube, Matthias O Franz, and Georg Umlauf. Learnt knot placement in b-spline curve approximation using support vector machines. *Computer Aided Geometric Design*, pages 104–116, 2018.

[55] Pascal Laube, Michael Grunwald, Matthias O Franz, and Georg Umlauf. Image inpainting for high-resolution textures using cnn texture synthesis. In *EG UK Computer Graphics & Visual Computing (2018)*. EGUK, 2018.

[56] Pascal Laube and Georg Umlauf. A short survey on recent methods for cage computation. In *Symposium on Information and Communication Systems BW-CAR— SINCOM (2016)*, pages 37–42, 2016.

[57] Pascal Laube, Georg Umlauf, and M. O. Franz. Evaluation of features for svm-based classification of geometric primitives in point clouds. In *IAPR International Conference on Machine Vision Applications*, pages 59–62. IEEE, 2017.

[58] Yann LeCun, Léon Bottou, Yoshua Bengio, and Patrick Haffner. Gradient-based learning applied to document recognition. *Proceedings of the IEEE*, 86(11):2278–2324, 1998.

[59] Eugene TY Lee. Choosing nodes in parametric curve interpolation. *Computer-Aided Design*, 21(6):363–370, 1989.

[60] Chuan Li and Michael Wand. Combining markov random fields and convolutional neural networks for image synthesis. In *Proceedings of the IEEE Conference on Computer Vision and Pattern Recognition*, pages 2479–2486, 2016.

[61] Weishi Li, Shuhong Xu, Gang Zhao, and Li Ping Goh. Adaptive knot placement in b-spline curve approximation. *Computer-Aided Design*, 37(8):791–797, 2005.

[62] Xin Li, Jianmin Zheng, Thomas W Sederberg, Thomas JR Hughes, and Michael A Scott. On linear independence of t-spline blending functions. *Computer Aided Geometric Design*, 29(1):63–76, 2012.

[63] Choong-Gyoo Lim. A universal parametrization in b-spline curve and surface interpolation. *Computer Aided Geometric Design*, 16(5):407–422, 1999.

[64] Or Litany, Alex Bronstein, Michael Bronstein, and Ameesh Makadia. Deformable shape completion with graph convolutional autoencoders. *arXiv preprint arXiv:1712.00268*, 2017.

[65] Gang Liu, Yann Gousseau, and Gui-Song Xia. Texture synthesis through convolutional neural networks and spectrum constraints. In *Pattern Recognition (ICPR), 2016 23rd International Conference on*, pages 3234–3239. IEEE, 2016.

[66] Weiyin Ma and Jean-Pierre Kruth. Parameterization of randomly measured points for least squares fitting of b-spline curves and surfaces. *Computer-Aided Design*, 27(9):663–675, 1995.

[67] Daniel Maturana and Sebastian Scherer. Voxnet: A 3d convolutional neural network for real-time object recognition.

In *Intelligent Robots and Systems (IROS), 2015 IEEE/RSJ International Conference on*, pages 922–928. IEEE, 2015.

[68] R. Osada, T. Funkhouser, B. Chazelle, and D. Dobkin. Shape distributions. *ACM Trans. on Graphics*, 21(4):807–832, 2002.

[69] Chavdar Papazov and Darius Burschka. An efficient ransac for 3d object recognition in noisy and occluded scenes. In *Asian Conference on Computer Vision*, pages 135–148. Springer, 2010.

[70] HT Park, MH Chang, and SC Park. A slicing algorithm of point cloud for rapid prototyping. In *Proceedings of the 2007 Summer Computer Simulation Conference*, page 24. Society for Computer Simulation International, 2007.

[71] Hyungjun Park and Joo-Haeng Lee. B-spline curve fitting based on adaptive curve refinement using dominant points. *Computer-Aided Design*, 39(6):439–451, 2007.

[72] Deepak Pathak, Philipp Krahenbuhl, Jeff Donahue, Trevor Darrell, and Alexei A Efros. Context encoders: Feature learning by inpainting. In *Proceedings of the IEEE Conference on Computer Vision and Pattern Recognition*, pages 2536–2544, 2016.

[73] Les Piegl and Wayne Tiller. *The NURBS book*. Springer Science & Business Media, 2012.

[74] Les A Piegl and Wayne Tiller. Computing the derivative of nurbs with respect to a knot. *Computer aided geometric design*, 15(9):925–934, 1998.

[75] Les A Piegl and Wayne Tiller. Surface approximation to scanned data. *The Visual Computer*, 16(7):386–395, 2000.

[76] John Platt et al. Probabilistic outputs for support vector machines and comparisons to regularized likelihood methods. *Advances in large margin classifiers*, 10(3):61–74, 1999.

[77] Javier Portilla and Eero P Simoncelli. A parametric texture model based on joint statistics of complex wavelet coefficients. *International journal of computer vision*, 40(1):49–70, 2000.

[78] Hartmut Prautzsch, Wolfgang Boehm, and Marco Paluszny. *Bézier and B-spline techniques*. Springer Science & Business Media, 2013.

[79] Lutz Prechelt. Early stopping-but when? In *Neural Networks: Tricks of the trade*, pages 55–69. Springer, 1998.

[80] Charles R Qi, Hao Su, Kaichun Mo, and Leonidas J Guibas. Pointnet: Deep learning on point sets for 3d classification and segmentation. *Proc. Computer Vision and Pattern Recognition (CVPR), IEEE*, 1(2):4, 2017.

[81] Anshuman Razdan. Knot placement for b-spline curve approximation. *Tempe, AZ: Arizona State University*, 1999.

[82] Frank Rosenblatt. The perceptron: a probabilistic model for information storage and organization in the brain. *Psychological review*, 65(6):386, 1958.

[83] Michaël Roy, Sebti Foufou, Andreas Koschan, Frédéric Truchetet, and Mongi Abidi. Multiresolution analysis for irregular meshes. In *Wavelet Applications in Industrial Processing*, volume 5266, pages 249–260. International Society for Optics and Photonics, 2004.

[84] R. B. Rusu, Z. C. Marton, N. Blodow, and M. Beetz. Persistent point feature histograms for 3d point clouds. In *Int. Conf. on Intelligent Autonomous Systems*, 2008.

[85] R.B. Rusu and S. Cousins. 3D is here: Point Cloud Library (PCL). In *Int. Conf. on Robotics and Automation*, 2011.

[86] Muhammad Sarfraz and Syed Arshad Raza. Capturing outline of fonts using genetic algorithm and splines. In *Information*

Visualisation, 2001. Proceedings. Fifth International Conference on, pages 738–743. IEEE, 2001.

[87] Biplab Sarkar and Chia-Hsiang Menq. Parameter optimization in approximating curves and surfaces to measurement data. *Computer Aided Geometric Design*, 8(4):267–290, 1991.

[88] Oliver Schall, Alexander Belyaev, and Hans-Peter Seidel. Robust filtering of noisy scattered point data. In *Eurographics/IEEE VGTC Symposium Proceedings Point-Based Graphics*, pages 71–144. IEEE, 2005.

[89] Ruwen Schnabel, Roland Wahl, and Reinhard Klein. Efficient ransac for point-cloud shape detection. In *Computer graphics forum*, volume 26, pages 214–226. Wiley Online Library, 2007.

[90] Thomas W Sederberg, Jianmin Zheng, Almaz Bakenov, and Ahmad Nasri. T-splines and T-NURCCs. *ACM Transactions on Graphics*, 22(3):477–484, 2003.

[91] S Mariyam Hj Shamsuddin and Mahmoud Ali Ahmed. A hybrid parameterization method for nurbs. In *Computer Graphics, Imaging and Visualization, 2004. CGIV 2004. Proceedings. International Conference on*, pages 15–20. IEEE, 2004.

[92] Karen Simonyan and Andrew Zisserman. Very deep convolutional networks for large-scale image recognition. *arXiv preprint arXiv:1409.1556*, 2014.

[93] Nitish Srivastava, Geoffrey Hinton, Alex Krizhevsky, Ilya Sutskever, and Ruslan Salakhutdinov. Dropout: a simple way to prevent neural networks from overfitting. *The Journal of Machine Learning Research*, 15(1):1929–1958, 2014.

[94] Chen Sun, Abhinav Shrivastava, Saurabh Singh, and Abhinav Gupta. Revisiting unreasonable effectiveness of data in deep learning era. In *2017 IEEE International Conference on Computer Vision (ICCV)*, pages 843–852. IEEE, 2017.

[95] Gabriel Taubin. Estimation of planar curves, surfaces, and nonplanar space curves defined by implicit equations with applications to edge and range image segmentation. *IEEE Transactions on Pattern Analysis & Machine Intelligence*, 13(11):1115–1138, 1991.

[96] Theano Development Team. Theano: A Python framework for fast computation of mathematical expressions. *arXiv e-prints*, abs/1605.02688, May 2016.

[97] Leonard Thießen, Pascal Laube, Georg Umlauf, and Matthias Franz. Merging multiple 3d face reconstructions. In *Symposium on Information and Communication Systems BW-CAR— SINCOM (2014)*, volume 2014, pages 7–12, 2014.

[98] Vahit Tongur and Erkan Ülker. B-spline curve knot estimation by using niched pareto genetic algorithm (npga). In *Intelligent and Evolutionary Systems*, pages 305–316. Springer, 2016.

[99] Erkan Ülker. B-spline curve approximation using pareto envelope-based selection algorithm-pesa. *International Journal of Computer and Communication Engineering*, 2(1):60, 2013.

[100] Olga Valenzuela, Blanca Delgado-Marquez, and Miguel Pasadas. Evolutionary computation for optimal knots allocation in smoothing splines. *Applied Mathematical Modelling*, 37(8):5851–5863, 2013.

[101] V Vapnik. Statistical learning theory new york. *NY: Wiley*, 1998.

[102] Pascal Vincent, Hugo Larochelle, Isabelle Lajoie, Yoshua Bengio, and Pierre-Antoine Manzagol. Stacked denoising autoencoders: Learning useful representations in a deep network with a local denoising criterion. *Journal of machine learning research*, 11(Dec):3371–3408, 2010.

[103] E. Wahl, U. Hillenbrand, and G. Hirzinger. Surflet-pair-relation histograms: A statistical 3d-shape representation for rapid classification. In *3DIM*, pages 474–482, 2003.

[104] Thomas SA Wallis, Christina M Funke, Alexander S Ecker, Leon A Gatys, Felix A Wichmann, and Matthias Bethge. A parametric texture model based on deep convolutional features closely matches texture appearance for humans. *Journal of Vision*, 17(12):5–5, 2017.

[105] Zhou Wang, Alan C Bovik, Hamid R Sheikh, and Eero P Simoncelli. Image quality assessment: from error visibility to structural similarity. *IEEE transactions on image processing*, 13(4):600–612, 2004.

[106] Yonatan Wexler, Eli Shechtman, and Michal Irani. Space-time completion of video. *IEEE Transactions on pattern analysis and machine intelligence*, 29(3), 2007.

[107] YF Wu, YS Wong, Han Tong Loh, and YF Zhang. Modelling cloud data using an adaptive slicing approach. *Computer-Aided Design*, 36(3):231–240, 2004.

[108] Zhirong Wu, Shuran Song, Aditya Khosla, Fisher Yu, Linguang Zhang, Xiaoou Tang, and Jianxiong Xiao. 3d shapenets: A deep representation for volumetric shapes. In *Proceedings of the IEEE conference on computer vision and pattern recognition*, pages 1912–1920, 2015.

[109] Akihiro Yamamoto, Masaki Tezuka, Toshiya Shimizu, and Ryutarou Ohbuchi. Shrec'08 entry: Semi-supervised learning for semantic 3d model retrieval. In *Shape Modeling and Applications, 2008. SMI 2008. IEEE International Conference on*, pages 241–243. IEEE, 2008.

[110] Chao Yang, Xin Lu, Zhe Lin, Eli Shechtman, Oliver Wang, and Hao Li. High-resolution image inpainting using multi-scale neural patch synthesis. *arXiv preprint arXiv:1611.09969*, 2016.

[111] Xunnian Yang and Jianmin Zheng. Approximate t-spline surface skinning. *Computer-Aided Design*, 44(12):1269–1276, 2012.

[112] Fujiichi Yoshimoto, Masamitsu Moriyama, and Toshinobu Harada. Automatic knot placement by a genetic algorithm for data fitting with a spline. In *Proceedings of International Conference on Shape Modeling and Applications*, pages 162–169. IEEE, 1999.

[113] Yuan Yuan, Nan Chen, and Shiyu Zhou. Adaptive b-spline knot selection using multi-resolution basis set. *IIE Transactions*, 45(12):1263–1277, 2013.

[114] Xiuyang Zhao, Caiming Zhang, Bo Yang, and Pingping Li. Adaptive knot placement using a gmm-based continuous optimization algorithm in b-spline curve approximation. *Computer-Aided Design*, 43(6):598–604, 2011.

[115] Ciyou Zhu, Richard H Byrd, Peihuang Lu, and Jorge Nocedal. Algorithm 778: L-bfgs-b: Fortran subroutines for large-scale bound-constrained optimization. *ACM Transactions on Mathematical Software (TOMS)*, 23(4):550–560, 1997.

A Appendix

A.1 Runtime Remarks Section 3.2.4

Training and evaluation were parallelized on two systems with Intel-Core-i7@3, 4GHz 4-core processors, each system with 16GB RAM. With this setup, the training data generation took 51 hours while grid-search and evaluation of trained SVMs took 42 hours. We used our complete set of 14 features for training and evaluations and did not test subsets of these features due to the SVMs ability to discover meaningful and separable representations in high dimensional space. All methods were implemented using MATLAB.

A.2 Runtime Remarks Section 5.3

Inpainting results where computed using an Nvidia GeForce 1080 Ti GPU. The computation, excluding the extraction of Gramians for patch look-up, took roughly 8 min strongly depending on the number of iterations of the L-BFGS-B [115] optimization. The method was implemented using Theano [96].

© Springer Fachmedien Wiesbaden GmbH, part of Springer Nature 2020
P. Laube, *Machine Learning Methods for Reverse Engineering of Defective Structured Surfaces*, Schriftenreihe der Institute für Systemdynamik (ISD) und optische Systeme (IOS), https://doi.org/10.1007/978-3-658-29017-7

Printed in the United States
By Bookmasters